T0140383

Machine Learning: Foundations, Methodologies, and Applications

Series Editors

Kay Chen Tan, Department of Computing, Hong Kong Polytechnic University, Hong Kong, China

Dacheng Tao, University of Technology, Sydney, Australia

Books published in this series focus on the theory and computational foundations, advanced methodologies and practical applications of machine learning, ideally combining mathematically rigorous treatments of a contemporary topics in machine learning with specific illustrations in relevant algorithm designs and demonstrations in real-world applications. The intended readership includes research students and researchers in computer science, computer engineering, electrical engineering, data science, and related areas seeking a convenient medium to track the progresses made in the foundations, methodologies, and applications of machine learning.

Topics considered include all areas of machine learning, including but not limited to:

- Decision tree
- Artificial neural networks
- Kernel learning
- Bayesian learning
- Ensemble methods
- Dimension reduction and metric learning
- Reinforcement learning
- Meta learning and learning to learn
- Imitation learning
- Computational learning theory
- Probabilistic graphical models
- Transfer learning
- Multi-view and multi-task learning
- Graph neural networks
- Generative adversarial networks
- Federated learning

This series includes monographs, introductory and advanced textbooks, and state-of-the-art collections. Furthermore, it supports Open Access publication mode.

Davide Pastorello

Concise Guide to Quantum Machine Learning

 Springer

Davide Pastorello
Department of Information Engineering
and Computer Science
University of Trento
Trento, Italy

Trento Institute for Fundamental Physics
and Applications (TIFPA-INFN)
Trento, Italy

ISSN 2730-9908 ISSN 2730-9916 (electronic)
Machine Learning: Foundations, Methodologies, and Applications
ISBN 978-981-19-6899-0 ISBN 978-981-19-6897-6 (eBook)
https://doi.org/10.1007/978-981-19-6897-6

This Springer imprint is published by the registered company Springer Nature Singapore Pte Ltd.
The registered company address is: 152 Beach Road, #21-01/04 Gateway East, Singapore
189721, Singapore

To Chiara, Matilde, and Ludovica

Preface

This book is based on the lecture notes of the course *Quantum Machine Learning* for the Master's degree in Computer Science that I gave at the University of Trento in the academic years 2020-2021 and 2021-2022. The main purpose of the volume is to present a compact but effective introduction to the basics of quantum machine learning (QML) starting from fundamental notions of quantum mechanics and quantum computing. The approach followed is bottom-up and aimed at explaining how the foundations of quantum mechanics enable new and efficient learning schemes to readers with no background in quantum mechanics and quantum computing. In particular, the book provides an essential toolkit of mathematical concepts that are crucial for the required quantum formalism, while avoiding unnecessary details. Further, the content is not divided into a "classical part" that describes standard machine learning schemes and a "quantum part" that addresses their quantum counterparts. Instead, we adopt a slightly different approach in order to immerse the reader in the quantum realm from the outset. I want to clarify right now that QML is not merely the translation of classical[2] machine learning techniques into the language of quantum computing, but rather a new approach to data representation and processing that is intrinsically different from that performed by standard computers. Moreover, the quantum paradigms are also relevant in developing new classical machine learning algorithms within the domain of so-called *quantum-inspired* machine learning. Thus, the interplay between QML and classical machine learning deserves serious consideration also for providing hybrid quantum-classical architectures that can improve the existing ones. More generally, the notion that quantum computers will replace traditional computers at all is misguided; most likely, it will never happen. Classical computers are tremendously reliable and robust machines and the aim of quantum computing is not to supplant them.

To gain the most from this book, the reader should essentially be familiar with basic concepts of statistics and probability theory, and fundamentals of linear algebra like: vector spaces and their bases, linear maps, matrix operations, eigenvalues and

---

[2]In this book, *classical* means *non-quantum*, as in physics.

eigenvectors, diagonalizability, tensor products. Knowledge of quantum computing and machine learning is obviously an excellent starting point; nevertheless, I have sought to present the foundations of QML in a self-contained manner.

I am grateful to some people for useful discussions about the topics of the book and/or for suggestions to improve the manuscript. I would like to thank E. Blanzieri, E. Zardini, R. Leporini, A. Di Pierro, A. Rumyantsev, V. Cavecchia, P. Hauke, L. Schmid, D. Vinci. I am grateful to the Springer editor M. Sugino for the highly professional editorial process.

This work was supported by Q@TN, the joint lab of the University of Trento, FBK - Fondazione Bruno Kessler, INFN - National Institute for Nuclear Physics, and CNR - National Research Council.

University of Trento. August 30, 2022   Davide Pastorello

Contents

Chapter 1

Introduction

Quantum machine learning (QML) is a rapidly emerging research area where the power of quantum computing is applied to machine learning tasks in order to pursue computational advantages in view of the present context of ever-growing amounts of data to manage. More generally, QML turns out be a promising path to explore new opportunities offered by quantum machines, both the existing ones and those not yet built. It is currently a hot topic within the scientific communities working on quantum technologies and attracts interests also from industries and non-academic institutions which are excited by the idea of dealing with big data running machine learning algorithms on quantum computers. This volume essentially provides an introduction to the possibilities offered by quantum computers to machine learning discussing the fundamental aspects and presenting an overview on the most recent and remarkable developments in this scientific and technological area.

Data encoding into quantum states and information processing based on quantum phenomena like *quantum superposition* and *quantum entanglement* can be used to formulate machine learning schemes within the well established framework of quantum computing. QML algorithms may present relevant advantages with respect to classical machine learning algorithms in terms of time and space complexity. Remarkable examples are given by the embedding of quantum subroutines (such as Qdist, see section 5.4) into machine learning schemes to efficiently calculate distances in the feature space with advantages in classification and clustering or Grover-based subroutines to find an item in an unsorted database, with a quadratic speedup with respect to exhaustive search, applied to pattern recognition. The first ideas on quantum versions of machine learning algorithms were presented about twenty years ago (e.g. [VM00, Tr01, Sc03]) but the real interest in QML has sparked in the last decade around the development of the first working prototypes of quantum machines like those manufactured by IBM [IBM], Rigetti [Rig], and D-Wave

D. Pastorello, *Concise Guide to Quantum Machine Learning*,
Machine Learning: Foundations, Methodologies, and Applications,
https://doi.org/10.1007/978-981-19-6897-6_1

Systems [DW21a] and the publication of many theoretical and experimental results (e.g. [Re14, Wi15, Ro17, Sc18, Vo19, Sa19, Ta20, Je21, Be20, Ab21, Sg21]).

1.1 Machine learning generalities

Machine learning is the science of making computers able to accomplish tasks extracting information from data without being explicitly programmed. Machine learning has revolutionized the way in which data are processed. Today, machine learning models are developed for a huge amount of tasks such as medical diagnoses, fraud detection, predictions in finance and marketing. The success of machine learning in many disciplines is mainly due to the recent availability of relatively powerful computers. The general goal of a machine learning algorithm is to reveal insights into an observed phenomenon, building a mathematical model, on the basis of collected data instances. In this regard, heuristic properties of the phenomenon are called *features*, typically expressed by numerical values. For example, numerical features may be medical parameters of human beings in a considered population and a machine learning algorithm can be used to predict whether an unknown person has a heart disease. Data instances can be represented by real vectors in a feature space, so if d is the number of considered features then the feature space identifies with $\mathbb{R}^d$, specifically each feature is associated to a vector of the standard basis of $\mathbb{R}^d$, assuming that the features are mutually independent, so the entries of a data vector correspond to the values of the features. Correlations between features may destroy the efficiency of machine learning algorithms, so there are techniques to choose effective features like *feature selection* (removing redundant features) and *feature extraction* (definition of new effective features from the existing ones).

An important notion in machine learning is a well-defined distance in the considered feature space which provides a metric structure where one can find patterns, clusters, or other geometric properties of data. Popular examples of distances are the *Euclidean distance* in $\mathbb{R}^d$ for real features:

$$d(\mathbf{x}_1, \mathbf{x}_2) = \sqrt{\sum_{j=1}^{d}(x_{1j} - x_{2j})^2}, \qquad (1.1.1)$$

where x_{ij} is the jth component of the data vector $\mathbf{x}_i$, and the *Hamming distance* in $\{0,1\}^d$ for binary features:

$$d(\mathbf{x}_1, \mathbf{x}_2) = \sum_{j=1}^{d} x_{1j} \oplus x_{2j}, \qquad (1.1.2)$$

where $\oplus$ is the sum modulo 2.

In the context of *supervised learning*, data points in the feature space $\mathcal{F}$ are labelled. Given a set of labelled data, the task of a supervised learning algorithm is to predict the correct labels of previously unseen instances. If Y is a set of numerical labels (continuous, discrete, or binary) the goal of supervised learning is approximating a function $F : \mathcal{F} \to Y$ given a *training set* $\{(\mathbf{x}_1, F(\mathbf{x}_1)), ..., (\mathbf{x}_N, F(\mathbf{x}_N))\}$ that is a collection of points in $\mathcal{F}$ with the corresponding values of F. An approximation of F can be determined by means of a *predictive model* that is a function $f : \mathcal{F} \to Y$ whose values are called *predictions*. Within parametric methods, a model f depends on a set of parameters θ so that we have to estimate the optimal parameters that provide a good approximation of F allowing the generalization of the model. The parameter estimation, or the *model training*, can be done through the optimization of a *loss function* like the following:

$$\mathcal{L}(\theta) := \sum_{i=1}^{N} |f(\mathbf{x}_i, \theta) - y_i|^2,$$

so that the global minimum of $\mathcal{L}$ can be searched, for example, by gradient descent. After the training, the model provides a prediction of the label y for a previously unseen data instance $\mathbf{x}$:

$$y = f(\mathbf{x}, \hat{\theta}) \quad \text{with} \quad \hat{\theta} = \mathrm{argmin}\mathcal{L}(\theta).$$

In *unsupervised learning*, there are no labels and algorithms must extract information from data automatically by recognition of structures in the considered dataset. For example, clustering algorithms find clusters of data instances such that the elements of each cluster satisfy a requirement of mutual similarity. Assume we have a dataset represented in a feature space, and we need to determine a partition of the dataset into clusters of similar data. A general algorithm for clustering is the following:

1. Choose an initial partition of the dataset into K clusters;

2. Calculate the centroid of each cluster;

3. Attach any data instance to the closets centroid (new partition);

4. Iterate until no new partition is found.

The third macro-area of machine learning is *reinforcement learning* in which an agent gets rewarded or punished according to a given rule for its decisions learning an optimal strategy by trial and error. This book focuses on supervised and unsupervised QML schemes; however, also quantum versions of reinforcement learning have been proposed [Do08, Je21, Sg21].

Nowadays, the success of machine learning starts to be partially undermined by some limitations of the current hardware. Many machine learning models can be very computationally demanding, sometimes requiring hours, days, or even months of training time on classical computers. Moreover, conventional computers are beginning to approach physical limitations that probably will slow their improvements in the next years. Among the possible alternative computing platforms to train and execute machine learning models, quantum computers represent a particularly interesting candidate.

1.2   Machine learning with quantum computers

Historically speaking, the research area of machine learning was born in the fifties from several proposals such as the Turing's learning machine [Tu50] and the perceptron [Ro58]. On the other hand, the field of quantum computing was established in the early eighties [Be80, Fe82]. So, why is QML such a recent research area? To answer this question we have to take into account that the availability of working prototypes of quantum machines was raised in the last years allowing the first interesting experiments running quantum algorithms. Moreover, the quantum algorithms which have been developed since the nineties still waiting to be definitely applied to accomplish concrete tasks and to solve real-life problems. Devising quantum algorithms for machine learning is currently considered a promising (maybe the most promising) path toward an effective impact of quantum computing outside the physical sciences.

Algorithm	Quantum tool	Quantum speedup
Principal component analysis	Phase estimation	Exponential
K-medians	Grover + Qdist	Quadratic
Hierarchical clustering	Grover + Qdist	Quadratic
K-means	Qdist	Exponential
Several classifiers	Qdist	Exponential
Support vector machines	HHL algorithm	Exponential
Neural networks	Variational circuits	?

Table 1.1. Some machine learning algorithms for which quantum versions exist with corresponding "quantum tools" and speedups with respect to the classical versions [Aï13, Ll13, Ll14, Re14, Wi14, Sc17, Sc18, PB21]. The tools are: Grover's algorithm for searching

in an unsorted database, the Qdist routine to calculate Euclidean distances, the phase estimation algorithm for eigendecomposition, the HHL algorithm to solve linear systems, variational quantum circuits depending on free parameters.

In the introductory chapters on quantum computing and quantum algorithms we review the fundamental aspects which are behind the so-called quantum speedups. As an example of the QML algorithms that we are going to consider, an essential summary of the quantum versions of celebrated machine learning algorithms, with corresponding quantum speedups, is reported in Table 1.1. Nevertheless, speedup is not the only quantum advantage which is pursued in QML, also the improvement of space efficiency, accuracy, and expressive power is a desirable goal in using quantum computers. In particular, quantum neural networks represent a research field under intense scientific investigation, however there are not well-established theoretical results or empirical evidences proving the quantum advantage of quantum neural networks over the classical counterpart like for other models. A quantum advantage of implementing neural networks on quantum computers is difficult to prove because the power of neural networks is well-known in practice and typically demonstrated in terms of benchmarks but there is the lack of satisfactory theoretical descriptions. On the contrary, the theoretical analysis is currently the more effective strategy to estimate the performances of quantum algorithms which, for now, can hardly be tested in trials of practical relevance. However, some influential experts argued that the quantum advantage may be not the right goal for QML [Sc22] suggesting that this vision forces the focus to a small class of problems hindering an advisable general and foundational approach. Therefore, QML must not be considered uniquely as quantum-enhanced machine learning but a new approach to data representation and processing which, for instance, can affect also classical machine learning by the development of algorithms within the so-called *quantum-inspired* machine learning [Se19, LP22].

1.3 Structure of the book

This volume is organized as follows: In chapters 2 and 3, we give a brief introduction to fundamentals of quantum mechanics and foundations of quantum computing (these parts can be essentially skipped by readers with sufficient knowledge of quantum mechanics and quantum computing). In turn, we discuss how to encode classical data into quantum states as well as the notions of quantum gates and quantum circuits in a very compact exposition using a standard notation. Moreover, we introduce the idea of adiabatic quantum computing since some of the presented QML schemes are formulated within this model. In chapter 4, we review the most relevant quantum algorithms for machine learning applications like the quantum Fourier transform that is a quantum algorithm to efficiently calculate the discrete Fourier transform, the Grover's search algorithm and the phase estimation algorithm that are important subroutines for QML algorithms. Chapter 5 provides a sort of *QML toolkit* including some crucial notions which are the basis for many QML schemes like the quantum random access memory for data retrieval and the routine Qdist to efficiently calculate Euclidean distances in a feature space. In chapter 6, we introduce QML algorithms for clustering that represent the quantum versions of well-known classical machine learning schemes such as K-means, K-medians, divisive clustering. In chapter 7, we overview some recently proposed quantum classifiers, the quantum implementation of a support vector machine, and an example of quantum-inspired machine learning defining a classical algorithm for classification based on a representation of data that is given in terms of the mathematical objects of the quantum formalism. Chapter 8 is mainly devoted to describing a modification of the Grover's algorithm to realize an associative memory; moreover, we summarize an adiabatic prescription and an application of the quantum Fourier transform for pattern recognition. In chapter 9, we introduce some quantum implementations of the perceptron model and the notion of quantum neural networks in terms of a composition of quantum perceptrons and in terms of variational quantum circuits. Moreover, quantum Boltzmann machines, quantum convolutional networks and quantum generative adversarial networks are also introduced. In the last chapter, there are some conclusions about the general aspects of QML as a research area that is under the spotlight of academy and industry trying to point out those aspects that are not clarified yet and deserve further investigation in the next years in order to find out better the possibilities offered by quantum machines.

Chapter 2

Basics of Quantum Mechanics

Quantum mechanics is the theory used to describe the physical objects of the "microscopic world" that are molecules, atoms, elementary particles such as electrons and photons. Classical mechanics based on the Newton's principles of dynamics and the classical electromagnetism based on Maxwell's equations do not work to describe microscopic physical systems. For instance, the classical electromagnetism does not provide the stability of the hydrogen atom that can be explained only in terms of its quantum description. In this chapter, we start from a list of general properties of the quantum physical systems that have been deduced as phenomenological evidences in quantum experiments. Next, we introduce the machinery of linear operators on Hilbert spaces to formalize the description of quantum systems.

2.1 Phenomenology

The general phenomenological evidences that characterize quantum systems can be formulated as the following properties:

1. *Randomness of measurement outcomes*: Repeated measurements of the *same* physical quantity (observable) A in the *same* physical conditions (state) produce *different* results. Let $\sigma(A) := \{a_1, ..., a_n\} \subset \mathbb{R}$ be a finite set of possible outcomes, repeated measurements of A, in the same state ψ, allow to construct a probability distribution $\mathbb{P}_\psi^{(A)}$ on $\sigma(A)$. For any A there exists a state ψ_a such that $\mathbb{P}_{\psi_a}^{(A)}(a) = 1$.

2. *Post-measurement state*: Let ψ be the physical state of the considered quantum system. If we perform a measurement process on the system to measure the observable A and the obtained outcome is $a \in \mathbb{R}$ then the state of the system, after the measurement, is ψ_a.

© The Author(s), under exclusive license to Springer Nature Singapore Pte Ltd. 2023
D. Pastorello, *Concise Guide to Quantum Machine Learning*,
Machine Learning: Foundations, Methodologies, and Applications,
https://doi.org/10.1007/978-981-19-6897-6_2

3. *Incompatible observables*: There are pairs of observables that cannot be simultaneously measured by an experiment.

The stochastic behavior of the experimental outcomes stated by property 1 is not
related to the finite precision of an instrument or to the uncertainty due to an incomplete knowledge of the state of the system. Property 1 refers to the *intrinsic*
uncertainty that does not depend on the observer but it is a fundamental characteristic of the physics of quantum systems. In this sense, in quantum mechanics
we drop the *realistic assumption*, that is, we do not assume that an observable
presents definite value before the measurement process. Property 2 corresponds to
the so-called *collapse of wave function*, it reads that in general a quantum measurement process alters the state of the observed system. In particular, in the resulting
post-measurement state the measured observable presents a definite value. Given
a quantum system, property 3 provides the existence of incompatible observables,
that is there is no an experimental apparatus that can be used to measure their
values simultaneously. On the one hand, if A and B are incompatible observables
and the state of the system is ψ_a then the outcome of a measurement process of B
is a random variable in $\sigma(B)$. On the other hand, if A and B are compatible then
there exists a state $\psi_{a,b}$ such that $\mathbb{P}^{(A)}_{\psi_{a,b}}(a) = \mathbb{P}^{(B)}_{\psi_{a,b}}(b) = 1$, that is, we can find a state
where both the observables have a definite value.

Example 2.1.1 *An electron presents an intrinsic angular momentum called* spin.
If one measures a spin component *along a fixed axis of a frame of reference, there
are two possible experimental outcomes:* $+\hbar/2$ *and* $-\hbar/2$, *where* $\hbar$ *is the* reduced
Planck constant[1]. *The value of a spin component can be measured in a* Stern-Gerlach
apparatus *by means of the interaction of the electron with an external magnetic field.
Let us denote the states corresponding to the values* $+\hbar/2$ *and* $-\hbar/2$ *by the symbols*
$|0\rangle$ *and* $|1\rangle$ *respectively. If the measurement outcome is* $\hbar/2$ $(-\hbar/2)$ *then the post-
measurement state of the electron is* $|0\rangle$ $(|1\rangle)$. *The unknown state* $|\psi\rangle$ *before the
measurement can be interpreted as a* superposition *of* $|0\rangle$ *and* $|1\rangle$:

$$|\psi\rangle = a|0\rangle + b|1\rangle, \tag{2.1.1}$$

where the coefficients a *and* b *are related to the probabilities of measuring* $+\hbar/2$
and $-\hbar/2$. *Moreover, the spin components along two different axes are incompatible
observables.*

In the next section, we introduce some notions that are the building blocks of the
mathematical structure of quantum mechanics where quantum phenomenology can

---

[1]$\hbar = 1.054571726 \times 10^{-34} J \cdot s$ as an SI unit but we typically set $\hbar$ to 1.

be described. In particular, the idea of superposition of physical states, which is related to the uncertainty over measurement outcomes, requires a structure of vector space in order to be mathematically founded.

2.2 Mathematical framework

The mathematical formulation of quantum mechanics is probably one of the most rich and complicated theoretical constructions ever built. In general the mathematical description of a quantum system is provided in terms of linear operators acting on separable infinite-dimensional Hilbert spaces. However, we just deal with finite-dimensional spaces in view of the main scope of this volume that is not a comprehensive treatise on quantum mechanics but an introduction to quantum computing and its application to machine learning. Therefore, fortunately, the required mathematical notions are just little more than elementary linear algebra which the reader can review in an infinity of textbooks if necessary.

Definition* 2.2.1**   *A (finite-dimensional)* ***Hilbert space *is a pair* $(\mathsf{H}, \langle \, | \, \rangle)$ *where* H *is a (finite-dimensional) complex vector space and* $\langle \, | \, \rangle : \mathsf{H} \times \mathsf{H} \to \mathbb{C}$ *is a map, called* ***inner product*** *on* H, *satisfying the following properties:*

   i) $\langle \, | \, \rangle$ *is linear in the right entry:*
      $\langle \psi | \alpha \varphi + \beta \phi \rangle = \alpha \langle \psi | \varphi \rangle + \beta \langle \psi | \phi \rangle \ \forall \psi, \varphi, \phi \in \mathsf{H} \ and \ \forall \alpha, \beta \in \mathbb{C};$

   ii) $\langle \, | \, \rangle$ *is conjugate symmetric:* $\langle \psi | \varphi \rangle = \langle \varphi | \psi \rangle^* \ \forall \psi, \varphi \in \mathsf{H}$, *where the symbol* *
      denotes the complex conjugation;

   iii) $\langle \, | \, \rangle$ *is positive-definite:* $\langle \psi | \psi \rangle > 0 \ \forall \psi \neq 0.$

A basis $\{\phi_i\}_i$ *of* H *is called* ***orthonormal basis*** *if* $\langle \phi_i | \phi_j \rangle = \delta_{ij}$, *where* δ_{ij} *is the Kronecker delta*[2].

The properties i) and ii) imply that the inner product is conjugate linear in the left entry: $\langle \alpha \psi + \beta \varphi | \phi \rangle = \alpha^* \langle \psi | \phi \rangle + \beta^* \langle \varphi | \phi \rangle \ \forall \psi, \varphi, \phi \in \mathsf{H}$ and $\forall \alpha, \beta \in \mathbb{C}$. One can easily verify that the inner product induces a norm $\| \cdot \|$ and a metric d on H by:

$$\|\psi\| := \sqrt{\langle \psi | \psi \rangle} \quad , \quad d(\psi, \varphi) := \|\psi - \varphi\| \qquad \psi, \varphi \in \mathsf{H}. \tag{2.2.1}$$

Let us just remark that the general definition of Hilbert space, including the infinite dimensional case, requires that H is a *complete space* with respect to the metric induced by the inner product. In the following, we focus on the finite-dimensional case and we usually denote the Hilbert space $(\mathsf{H}, \langle \, | \, \rangle)$ simply by H.

---

[2]The Kronecker delta is defined as $\delta_{ii} = 1$ and $\delta_{ij} = 0$ if $i \neq j$.

Definition 2.2.2 *Let* H *be a Hilbert space. A **linear operator on** H *is a map* $A : H \to H$ *satisfying the following property:*

$$A(\alpha\psi + \beta\varphi) = \alpha A\psi + \beta A\varphi, \tag{2.2.2}$$

for all $\alpha, \beta \in \mathbb{C}$ *and* $\psi, \varphi \in H$. *The set of linear operators on* H *is denoted by* $\mathfrak{B}(H)$. $A \in \mathfrak{B}(H)$ *is said to be **invertible** if there exists* $A^{-1} \in \mathfrak{B}(H)$, *called **inverse** of* A, *such that* $A^{-1}A = AA^{-1} = \mathbb{I}$ *where* $\mathbb{I}$ *is the identity operator on* H.

With the addition operation $(A + B)\psi := A\psi + B\psi$ and the scalar multiplication $(\alpha A)\psi := \alpha A\psi$, $\mathfrak{B}(H)$ is a complex vector space as well. The norm induced on a Hilbert space H by the inner product induces in turn a norm on $\mathfrak{B}(H)$, called *operator norm*, defined by:

$$\|A\|_{op} := \sup_{\|\psi\|=1} \|A\psi\|. \tag{2.2.3}$$

Definition 2.2.3 *Let* $\{\phi_i\}_i$ *be an orthonormal basis of* H *and* $A \in \mathfrak{B}(H)$. *The **trace** of* A *is defined by:*

$$\mathrm{tr}(A) := \sum_i \langle \phi_i | A\phi_i \rangle. \tag{2.2.4}$$

The definition of the trace does not depend on the choice of the basis. The map $A \mapsto \mathrm{tr}(A)$ is linear and invariant under cyclic permutations: $\mathrm{tr}(ABC) = \mathrm{tr}(CAB) = \mathrm{tr}(BCA)$.

   If H has dimension n, once fixed an orthonormal basis $\{\phi_i\}_{i=1,\dots,n}$ we can identify H to $\mathbb{C}^n$ and any linear operator on H can be represented by a $n \times n$ complex matrix:

$$A\psi = \begin{pmatrix} a_{11} & \cdot & \cdot & \cdot & a_{1n} \\ \cdot & \cdot & \cdot & \cdot & \cdot \\ \cdot & \cdot & \cdot & \cdot & \cdot \\ \cdot & \cdot & \cdot & \cdot & \cdot \\ a_{n1} & \cdot & \cdot & \cdot & a_{nn} \end{pmatrix} \begin{pmatrix} c_1 \\ \cdot \\ \cdot \\ \cdot \\ c_n \end{pmatrix}, \tag{2.2.5}$$

where the matrix elements of A and the components of ψ with respect to the basis $\{\phi_i\}_{i=1,\dots,n}$ are respectively given by: $a_{ij} = \langle \phi_i | A\phi_j \rangle$ and $c_i = \langle \phi_i | \psi \rangle$. Thus the trace of A is nothing but the sum of the diagonal elements $\mathrm{tr}(A) = \sum_{i=1}^n a_{ii}$. In coordinates, the composition of linear operators corresponds to the standard matrix multiplication.

Definition 2.2.4   *The **adjoint** of $A \in \mathfrak{B}(\mathsf{H})$ is the unique operator $A^{\dagger} \in \mathfrak{B}(\mathsf{H})$ such that:*

$$\langle A^{\dagger}\psi|\varphi\rangle = \langle\psi|A\varphi\rangle \qquad \forall\psi, \varphi \in \mathsf{H}.$$

*A is called **self-adjoint** if $A^{\dagger} = A$.*
*$U \in \mathfrak{B}(\mathsf{H})$ is called **unitary** if $U^{\dagger} = U^{-1}$.*

In the finite-dimensional context, in terms of matrices, the adjoint is nothing but the conjugate transpose and the self-adjoint operators correspond to hermitian matrices.

Definition 2.2.5   *The **spectrum** of $A \in \mathfrak{B}(\mathsf{H})$ is the set of its **eigenvalues**:*

$$\sigma(A) := \{\lambda \in \mathbb{C} : A\psi = \lambda\psi, \psi \in \mathsf{H} \setminus \{0\}\}. \tag{2.2.6}$$

*The **eigenspace** of $\lambda \in \sigma(A)$ is the subspace $\mathsf{H}_{\lambda} := \{\psi \in \mathsf{H} : A\psi = \lambda\psi\} \subset \mathsf{H}$. A nonzero vector $\psi \in \mathsf{H}_{\lambda}$ is called **eigenvector** of A with eigenvalue λ. The **spectral radius** of A is the positive number*

$$r(A) := \max_{\lambda \in \sigma(A)} |\lambda|. \tag{2.2.7}$$

Let us observe that the eigenvalues of a self-adjoint operator are real, in fact: let ψ be an eigenvector of the self-adjoint operator A with eigenvalue λ, thus:

$$\lambda\langle\psi|\psi\rangle = \langle\psi|A\psi\rangle = \langle A\psi|\psi\rangle = \lambda^{*}\langle\psi|\psi\rangle.$$

The eigenvalues of a unitary operator are complex numbers with unit modulus, also called *phases*: let φ be an eigenvector of the unitary operator U with eigenvalue μ, we have:

$$\langle\varphi|\varphi\rangle = \langle U\varphi|U\varphi\rangle = \mu^{*}\mu\langle\varphi|\varphi\rangle = |\mu|^{2}\langle\varphi|\varphi\rangle,$$

then $|\mu|^{2} = 1$, so $\mu = e^{i\phi}$ with $\phi \in \mathbb{R}$.

A nice property for a liner operator A on H is the existence of an orthonormal basis of H consisting of eigenvectors of A, in this case the operator is said to be *diagonalizable* because it is represented by a diagonal matrix with respect to the basis of its eigenvectors. The spectral theorem provides a complete characterization of diagonalizable operators.

Theorem 2.2.6 (Spectral theorem)   *$A \in \mathfrak{B}(\mathsf{H})$ is normal, that is, $AA^{\dagger} = A^{\dagger}A$ if and only if there exists an orthonormal basis of H made by eigenvectors of A.*

In particular, the spectral theorem implies that any self-adjoint operator is diagonalizable and admits a so-called spectral decomposition. Let H_{λ} be an eigenspace of the self-adjoint operator A. The orthogonal projector P_λ onto H_λ has the form:

$$P_\lambda := \sum_{i=1}^{\dim \mathsf{H}_\lambda} |\phi_i\rangle\langle\phi_i|, \tag{2.2.8}$$

where $\{\phi_i\}_{i=1,\ldots,\dim \mathsf{H}_\lambda}$ is an orthonormal basis of H_λ and the symbol $|\phi_i\rangle\langle\phi_i|$ denotes the orthogonal projector $P_i(\,\cdot\,) := |\phi_i\rangle\langle\phi_i|\,\cdot\,\rangle$ that projects onto the 1-dimensional subspace spanned by $|\phi_i\rangle$. The **spectral decomposition** of the self-adjoint operator A is:

$$A = \sum_{\lambda \in \sigma(A)} \lambda P_\lambda, \tag{2.2.9}$$

where $P_\lambda P_{\lambda'} = 0$ for $\lambda \neq \lambda'$. The collection of orthogonal projectors $\{P_\lambda\}_{\lambda \in \sigma(A)}$ is called **spectral measure** or **projection-valued measure (PVM)** of A. Given a map $f : \mathbb{R} \to \mathbb{R}$, we denote by $f(A)$ the self-adjoint operator whose eigenvalues are $f(\lambda)$ with $\lambda \in \sigma(A)$, then:

$$f(A) := \sum_{\lambda \in \sigma(A)} f(\lambda) P_\lambda. \tag{2.2.10}$$

Applying the spectral decomposition of a self-adjoint operator A, it is easy to check the equivalence between the operator norm and the spectral radius:

$$\|A\|_{op} = r(A). \tag{2.2.11}$$

As illustrated in the next section, in quantum mechanics, any physical quantity is described by a self-adjoint operator whose real spectrum represents the set of the possible outcomes of a measurement process of that quantity. The projectors of the associated spectral measure are the mathematical tools used for calculating quantum probabilities and describing the collapse of the quantum state due to the measurement process.

The notion of *tensor product* is crucial in quantum mechanics to describe *composite quantum systems*.

Definition 2.2.7 *Let H_A and H_B be Hilbert spaces and $\psi \in \mathsf{H}_A$, $\varphi \in \mathsf{H}_B$. The bilinear form $\psi \otimes \varphi : \mathsf{H}_A \times \mathsf{H}_B \to \mathbb{C}$ defined by:*

$$\psi \otimes \varphi(x, y) := \langle\psi|x\rangle_A \langle\varphi|y\rangle_B \qquad x \in \mathsf{H}_A, y \in \mathsf{H}_B, \tag{2.2.12}$$

*is called **tensor product of vectors** ψ and φ. The complex vector space:*

$$\mathsf{H}_A \otimes \mathsf{H}_B := \mathrm{span}\{\psi \otimes \varphi : \psi \in \mathsf{H}_A, \varphi \in \mathsf{H}_B\} \tag{2.2.13}$$

equipped with the inner product defined by:

$$\langle \psi \otimes \varphi | \psi' \otimes \varphi' \rangle := \langle \psi | \psi' \rangle_A \cdot \langle \varphi | \varphi' \rangle_B \qquad \psi, \psi' \in \mathsf{H}_A \, \varphi, \varphi' \in \mathsf{H}_B, \qquad (2.2.14)$$

*and extended by linearity, is called **tensor product of Hilbert spaces** H_A and H_B.*

*Given $A \in \mathfrak{B}(\mathsf{H}_A)$ and $B \in \mathfrak{B}(\mathsf{H}_B)$, the **tensor product of linear operators** A and B is:*

$$(A \otimes B)(\psi \otimes \varphi) := A\psi \otimes B\varphi \qquad (2.2.15)$$

that extends to $\mathsf{H}_A \otimes \mathsf{H}_B$ by linearity.

The complex vector space $\mathsf{H}_A \otimes \mathsf{H}_B$, equipped with the product defined in (2.2.14), is a Hilbert space as well. If $\dim \mathsf{H}_A = n$ and $\dim \mathsf{H}_B = m$ then $\dim(\mathsf{H}_A \otimes \mathsf{H}_B) = n \cdot m$. We define also the tensor product of the spaces of linear operators on H_A and H_B:

$$\mathfrak{B}(\mathsf{H}_A) \otimes \mathfrak{B}(\mathsf{H}_B) := \mathrm{span}\{A \otimes B : A \in \mathfrak{B}(\mathsf{H}_A), B \in \mathfrak{B}(\mathsf{H}_B)\}. \qquad (2.2.16)$$

We have the remarkable identity: $\mathfrak{B}(\mathsf{H}_A) \otimes \mathfrak{B}(\mathsf{H}_B) = \mathfrak{B}(\mathsf{H}_A \otimes \mathsf{H}_B)$. Therefore, any $T \in \mathfrak{B}(\mathsf{H}_A \otimes \mathsf{H}_B)$ can be written in the following form:

$$T = \sum_i A_i \otimes B_i, \qquad (2.2.17)$$

where $\{A_i\}_i \subset \mathfrak{B}(\mathsf{H}_A)$ and $\{B_i\}_i \subset \mathfrak{B}(\mathsf{H}_B)$.

Example 2.2.8   *If $\mathsf{H}_A = \mathbb{C}^n$ and $\mathsf{H}_B = \mathbb{C}^m$ then:*

$$A = \begin{pmatrix} a_{11} & \cdot & \cdot & \cdot & a_{1n} \\ \cdot & \cdot & \cdot & \cdot & \cdot \\ \cdot & \cdot & \cdot & \cdot & \cdot \\ \cdot & \cdot & \cdot & \cdot & \cdot \\ a_{n1} & \cdot & \cdot & \cdot & a_{nn} \end{pmatrix} \in \mathfrak{B}(\mathsf{H}_A) \qquad B = \begin{pmatrix} b_{11} & \cdot & \cdot & \cdot & b_{1m} \\ \cdot & \cdot & \cdot & \cdot & \cdot \\ \cdot & \cdot & \cdot & \cdot & \cdot \\ \cdot & \cdot & \cdot & \cdot & \cdot \\ b_{m1} & \cdot & \cdot & \cdot & b_{mm} \end{pmatrix} \in \mathfrak{B}(\mathsf{H}_B)$$

$$A \otimes B = \begin{pmatrix} a_{11}B & \cdot & \cdot & \cdot & a_{1n}B \\ \cdot & \cdot & \cdot & \cdot & \cdot \\ \cdot & \cdot & \cdot & \cdot & \cdot \\ \cdot & \cdot & \cdot & \cdot & \cdot \\ a_{n1}B & \cdot & \cdot & \cdot & a_{nn}B \end{pmatrix} \in \mathfrak{B}(\mathsf{H}_A \otimes \mathsf{H}_B)$$

The notion of partial trace is a generalization of the trace providing a sort of projection of an operator acting on a tensor product of Hilbert spaces, with general form (2.2.17), onto a tensor factor.

Definition 2.2.9 *The **partial trace** of $T \in \mathfrak{B}(\mathsf{H}_A \otimes \mathsf{H}_B)$ with respect to H_B is the unique operator $\mathrm{tr}_{\mathsf{H}_B}(T) \in \mathfrak{B}(\mathsf{H}_A)$ such that*

$$\mathrm{tr}\left[\mathrm{tr}_{\mathsf{H}_B}(T)A\right] = \mathrm{tr}[T(A \otimes \mathbb{I}_{\mathsf{H}_B})] \qquad \forall A \in \mathfrak{B}(\mathsf{H}_A). \qquad (2.2.18)$$

Let $\{\psi_i\}_i$ and $\{\varphi_j\}_j$ be orthonormal bases of H_A and H_B respectively, so $\{\psi_i \otimes \varphi_j\}_{ij}$ is an orthonormal basis of $\mathsf{H}_A \otimes \mathsf{H}_B$. The explicit computation of the partial trace of the operator $T \in \mathfrak{B}(\mathsf{H}_A \otimes \mathsf{H}_B)$ is done calculating the matrix elements:

$$[\mathrm{tr}_{\mathsf{H}_B}(T)]_{kl} = \sum_j \langle \psi_k \otimes \varphi_j | T \psi_l \otimes \varphi_j \rangle.$$

In the following, we will see that composite quantum systems are described in tensor product Hilbert spaces and the state of a single subsystem can be assigned by means of the partial trace of the *density operator* representing the state of the composite system.

According to the standard quantum formalism, in the following we adopt the *Dirac notation*: a vector in H with unit norm is denoted by the symbol $|\psi\rangle$, called *ket*, a vector of the *dual space* is denoted by the symbol $\langle\psi|$, called *bra*. In coordinates:

$$|\psi\rangle = \begin{pmatrix} c_1 \\ \vdots \\ c_n \end{pmatrix} \qquad \langle\psi| = (c_1^*, ..., c_n^*), \qquad (2.2.19)$$

where $c_i^* \in \mathbb{C}$ is the complex conjugate of $c_i \in \mathbb{C}$. The inner product $\langle\psi|\varphi\rangle$ of two vectors $|\psi\rangle$ and $|\varphi\rangle$ is also called *braket* and the projector P onto the 1-dimensional subspace spanned by $|\psi\rangle$ is written as $P = |\psi\rangle\langle\psi|$ as done in (2.2.8).

2.3   Quantum states and observables

We need to apply the mathematical tools introduced in the previous section to formalize the phenomenological evidences listed in section 2.1. The mathematical formulation of quantum mechanics is based on these postulates:

1. Any quantum system is associated to a Hilbert space H;

2. Observables are self-adjoint operators on H;

3. Pure states are equivalence classes of unit vectors in H with equivalence relation:

$$|\psi\rangle \sim |\varphi\rangle \Leftrightarrow \exists \alpha \in \mathbb{R} : |\psi\rangle = e^{i\alpha}|\varphi\rangle \quad \text{with} \quad |\psi\rangle, |\varphi\rangle \in \mathsf{H}. \tag{2.3.1}$$

In quantum mechanics, the physical quantities that can be measured in experiments on the considered quantum system are called *observables*. The expression *pure state* refers to a notion of quantum state that encode the maximum knowledge of the system physical conditions. On the other hand there is the notion of *mixed state*, introduced below, that provides a statistical description of the quantum system in presence of a lack of knowledge of its physical conditions. The pure state of a quantum system can be identified to a unit vector $|\psi\rangle \in \mathsf{H}$ with the requirement that any other unit vector differing by a multiplicative phase factor represent the same physical state (condition (2.3.1)). In other words, pure states can be identified with 1-dimensional orthogonal projectors in H that are the linear operators of the form $\rho_\psi = |\psi\rangle\langle\psi|$ with $\langle\psi|\psi\rangle = 1$. Thus when a physicist says that the state of a quantum system is $|\psi\rangle \in \mathsf{H}$, mathematically he means $|\psi\rangle\langle\psi|$.

If there is a lack of knowledge of the system physical conditions, its state is given by a statistical mixture of pure states, called *mixed state*:

$$\rho = \sum_i \lambda_i |\psi_i\rangle\langle\psi_i| \in \mathfrak{B}(\mathsf{H}), \tag{2.3.2}$$

where $\lambda_i \geq 0$ are statistical weights so that $\sum_i \lambda_i = 1$. The set of states of a quantum system described in the Hilbert space H is:

$$\mathfrak{S}(\mathsf{H}) = \{\rho \in \mathfrak{B}(\mathsf{H}) : \rho \geq 0, \operatorname{tr}(\rho) = 1\}, \tag{2.3.3}$$

where $\rho \geq 0$ means $\langle\psi|\rho\psi\rangle \geq 0$ for any $|\psi\rangle \in \mathsf{H}$. The set of quantum states is convex and the Krein-Millman theorem [KM40] entails that the elements of $\mathfrak{S}(\mathsf{H})$ are the 1-dimensional orthogonal projectors (pure states) and all their convex combinations (mixed states). The elements of $\mathfrak{S}(\mathsf{H})$ provide the general notion of quantum states in terms of normalized positive operators that are called *density matrices* or *density operators*.

Let $\{|\psi_i\rangle\}_i \subset \mathsf{H}$ be a finite set of pure states, the pure state defined by the normalized linear combination:

$$|\Psi\rangle := \frac{\sum_i a_i |\psi_i\rangle}{\| \sum_i a_i |\psi_i\rangle \|} \in \mathsf{H} \quad \text{with} \quad a_i \in \mathbb{C}, \tag{2.3.4}$$

is called **coherent superposition** of $\{|\psi_i\rangle\}_i$. Let $\{\rho_i\}_i \subset \mathfrak{S}(\mathsf{H})$ be a finite set of quantum states as density operators that can be pure or mixed, the operator defined by:

$$\rho := \sum_i \lambda_i \rho_i \quad \text{with} \quad \lambda_i \geq 0 \text{ and } \sum_i \lambda_i = 1, \tag{2.3.5}$$

is a density matrix called **incoherent superposition** of $\{\rho_i\}_i$. The coherent superposition (2.3.4) is allowed by the linear structure of the quantum theory and represents a physical property that completely deviates from classical physics, instead the incoherent superposition (2.3.5)is just a statistical mixture of states which is not directly related to the quantum nature of the described system.

Definition **2.3.1** *A quantum system described in a 2-dimensional Hilbert space is called* **qubit**.

Consider the following two pure states of a qubit that are identified to the vectors of the standard basis of $\mathbb{C}^2$:

$$|0\rangle = \begin{pmatrix} 1 \\ 0 \end{pmatrix} \qquad |1\rangle = \begin{pmatrix} 0 \\ 1 \end{pmatrix}. \tag{2.3.6}$$

We can consider a coherent superposition of $|0\rangle$ and $|1\rangle$:

$$|\psi\rangle = a|0\rangle + b|1\rangle \quad \text{with} \quad |a|^2 + |b|^2 = 1, \tag{2.3.7}$$

the pure state $|\psi\rangle$, as a projector, can be expressed in the matrix form:

$$|\psi\rangle\langle\psi| = \begin{pmatrix} |a|^2 & ab^* \\ ba^* & |b|^2 \end{pmatrix}. \tag{2.3.8}$$

An incoherent superposition of $|0\rangle$ and $|1\rangle$ is the mixed state:

$$\rho = \lambda_0|0\rangle\langle 0| + \lambda_1|1\rangle\langle 1| \quad \text{with} \quad \lambda_{0,1} \geq 0 \quad \text{and} \quad \lambda_0 + \lambda_1 = 1, \tag{2.3.9}$$

in matrix form:

$$\rho = \begin{pmatrix} \lambda_0 & 0 \\ 0 & \lambda_1 \end{pmatrix}. \tag{2.3.10}$$

If a system is prepared in a coherent superposition, for instance the superposition (2.3.8) of $|0\rangle$ and $|1\rangle$, its interaction with the environment can suppress the off-diagonal elements, then the system evolves into an incoherent superposition like

(2.3.10). This phenomenon is called **decoherence** and corresponds to an information loss of the initial state.

Now let us reconsider the three phenomenological evidences listed in section 2.1 in terms of the three postulates of the mathematical construction of quantum mechanics:

- *Randomness of measurement outcomes*: The possible experimental values of the observable A are the element of its spectrum $\sigma(A)$.

  Given a pure state $|\psi\rangle \in \mathsf{H}$, the probability of measuring the value $a \in \sigma(A)$ is:

  $$\mathbb{P}_\psi(a) = \langle\psi|P_a\psi\rangle, \tag{2.3.11}$$

  where $\{P_a\}_{a\in\sigma(A)}$ is the spectral measure of A. If we consider repeated measurements of the observable A in the same physical conditions represented by the state $|\psi\rangle$, the expectation value of A on the state $|\psi\rangle$ is the mean of the outcomes:

  $$\langle A\rangle_\psi := \sum_{a\in\sigma(A)} a\,\mathbb{P}_\psi(a) = \langle\psi|A\psi\rangle. \tag{2.3.12}$$

  In the case of a mixed state ρ we have the straightforward generalizations:

  $$\mathbb{P}_\rho(a) = \mathrm{tr}(P_a\rho) \quad \text{and} \quad \langle A\rangle_\rho = \mathrm{tr}(A\rho). \tag{2.3.13}$$

- *Post-measurement state*: Let $|\psi\rangle \in \mathsf{H}$ be the state of the considered quantum system. If we perform a measurement process of A with outcome $a \in \sigma(A)$, then the state of the system, after the measurement, is:

  $$|\psi_a\rangle = \frac{P_a|\psi\rangle}{\sqrt{\langle\psi|P_a\psi\rangle}}. \tag{2.3.14}$$

  In the case of a mixed state ρ, the post-measurement state is:

  $$\rho_a = \frac{P_a\rho P_a}{\mathrm{tr}(P_a\rho)}. \tag{2.3.15}$$

  Since the pure state (2.3.14) is an eigenvector of A, in quantum mechanics it is called *eigenstate* of the observable A.

- *Compatible and incompatible observables*: A and B are compatible when they commute:

$$[A, B] := AB - BA = 0, \tag{2.3.16}$$

in this case: $P_a^A P_b^B = P_b^B P_a^A$ $\forall a \in \sigma(A)$ and $\forall b \in \sigma(B)$, so the following probability is well-defined:

$$\mathbb{P}_\psi(A = a \wedge B = b) = \langle \psi | P_a^A P_b^B \psi \rangle = \langle \psi | P_b^B P_a^A \psi \rangle. \tag{2.3.17}$$

$\mathbb{P}_\psi(A = a \wedge B = b)$ is the joint probability of measuring the value a of the observable A *and* the value b of the observable B when the system is in the state $|\psi\rangle$. Conversely, if $[A, B] \neq 0$ then we have not a well-defined joint probability $\mathbb{P}_\psi(A = a \wedge B = b)$, this fact is consistent with the phenomenological evidence that A and B cannot be simultaneously measured. Moreover, let us remark that in the presented mathematical formulation of quantum mechanics the measurement process of the observable A is completely described by the PVM $\{P_a\}_{a \in \sigma(A)}$ which determines the probability distribution of the outcomes and the post-measurement state.

Example 2.3.2 *Let us reconsider the example 2.1.1. An electron admits a triple of observables called* components of spin (S_x, S_y, S_z). *If an electron is described in the frame of reference where it is at rest[3] then the associated Hilbert space is* $\mathsf{H} \simeq \mathbb{C}^2$. *The spin-operators are defined by:*

$$S_{x,y,z} := \frac{\hbar}{2} \sigma_{x,y,z} \, , \tag{2.3.18}$$

where $\sigma_{x,y,z}$ *are the Pauli matrices:*

$$\sigma_x = \begin{pmatrix} 0 & 1 \\ 1 & 0 \end{pmatrix}, \quad \sigma_y = \begin{pmatrix} 0 & -i \\ i & 0 \end{pmatrix}, \quad \sigma_z = \begin{pmatrix} 1 & 0 \\ 0 & -1 \end{pmatrix}. \tag{2.3.19}$$

If we measure the $\hat{z}$-component of the spin, the two possible outcomes are the eigenvalues of S_z: $+\frac{\hbar}{2}$ *and* $-\frac{\hbar}{2}$.
Since:

$$[\sigma_i, \sigma_j] = 2i\epsilon_{ijk}\sigma_k, \tag{2.3.20}$$

---

[3]In another frame of reference we must consider the kinetic degrees of freedom to specify the state of the electron, in that case we need an infinite-dimensional Hilbert space.

where ϵ_{ijk} is the Levi-Civita symbol*:*

$$\epsilon_{ijk} = \begin{cases} 1 & \text{if } (i,j,k) \text{ is an even permutation of } (x,y,z) \\ -1 & \text{if } (i,j,k) \text{ is an odd permutation of } (x,y,z) \\ 0 & \text{if there are repeated indexes} \end{cases} \quad (2.3.21)$$

then S_i and S_j are incompatible quantities when $i \neq j$. Pauli matrices are also used to describe the polarization of a photon. The states of vertical and horizontal polarization (with respect to the z-axis) of a photon are eigenstates of the observable rectilinear polarization described by the Pauli matrix σ_z:

$$|0\rangle = \begin{pmatrix} 1 \\ 0 \end{pmatrix} \qquad |1\rangle = \begin{pmatrix} 0 \\ 1 \end{pmatrix}. \quad (2.3.22)$$

Similarly diagonal polarization is described by the Pauli matrix σ_x, with eigenstates:

$$|+\rangle = \frac{|0\rangle + |1\rangle}{\sqrt{2}} \qquad |-\rangle = \frac{|0\rangle - |1\rangle}{\sqrt{2}}. \quad (2.3.23)$$

Finally circular polarization is described by Pauli matrix σ_y whose eigenvectors are:

$$|L\rangle = \frac{1}{\sqrt{2}} \begin{pmatrix} 1 \\ i \end{pmatrix} \qquad |R\rangle = \frac{1}{\sqrt{2}} \begin{pmatrix} 1 \\ -i \end{pmatrix} \quad (2.3.24)$$

$|L\rangle$ and $|R\rangle$ are called left polarization state *and* right polarization state *respectively.*

2.4 Quantum dynamics

The time evolution of an isolated[4] quantum system is mathematically described by a one-parameter group of unitary operators $\{U_t\}_{t \in \mathbb{R}}$ defined by:

$$U_t := \sum_{\lambda \in \sigma(\mathcal{H})} e^{-i\frac{t}{\hbar}\lambda} P_h \equiv e^{-i\frac{t}{\hbar}\mathcal{H}}, \quad (2.4.1)$$

where $\hbar$ is the reduced Planck constant, $\mathcal{H}$ is the Hamiltonian operator which represents the observable *total energy* of the considered system and $\{P_h\}_{h \in \sigma(\mathcal{H})}$ is the spectral measure of $\mathcal{H}$.

---

[4]The considered quantum system is assumed to be non-interacting with the environment. The notion of open system is introduced in the next section.

Example **2.4.1**   *Given an inertial frame of reference, the Hamiltonian of an electron at rest in a magnetic field* $\mathbf{B} = (B_x, B_y, B_z)$ *is:*

$$\mathcal{H} = -\gamma\, \boldsymbol{\sigma} \cdot \mathbf{B},$$

where $\gamma > 0$ *and* $\boldsymbol{\sigma} = (\sigma_x, \sigma_y, \sigma_z)$.

If the state at time $t = 0$ is $|\psi_0\rangle \in \mathsf{H}$ then the state at time $t > 0$ is:

$$|\psi_t\rangle = U_t|\psi_0\rangle = e^{-i\frac{t}{\hbar}\mathcal{H}}|\psi_0\rangle, \tag{2.4.2}$$

Taking the time derivative of equation (2.4.2), one obtains the *Schrödinger equation*:

$$i\hbar\frac{d}{dt}|\psi_t\rangle = \mathcal{H}|\psi_t\rangle, \tag{2.4.3}$$

that is the equation of motion of a quantum system with Hamiltonian $\mathcal{H}$. In case
of a time-dependent Hamiltonian, $\mathcal{H}$ must be replaced by a one-parameter family
of self-adjoint operators $\{\mathcal{H}(t)\}_{t\in\mathbb{R}}$ and the Schrödinger equation assumes the form:

$$i\hbar\frac{d}{dt}|\psi_t\rangle = \mathcal{H}(t)|\psi_t\rangle. \tag{2.4.4}$$

In particular, equation (2.4.4) is crucial to formulate the *adiabatic theorem* which is
the basis of Adiabatic Quantum Computing as discussed in section 3.4, providing a
universal model for quantum computation alternative to quantum circuits.

More generally, if the state of the system at $t = 0$ is a mixed state ρ_0, its time
evolution is described by:

$$\rho_t = U_t\rho_0 U_t^\dagger, \tag{2.4.5}$$

from which one obtains the equation of motion for mixed states:

$$i\hbar\frac{d}{dt}\rho_t = [\mathcal{H}, \rho_t], \tag{2.4.6}$$

called *Liouville-von Neumann equation*. If the initial state ρ_0 is a pure state then
(2.4.6) corresponds to (2.4.3).

2.5   Composite quantum systems

There are quantum systems, called *composite quantum systems*, presenting an inter-
nal structure so that one distinguishes in them two or more *subsystems* which can

be observed separately performing *local measurements*. Conversely, single quantum systems can be combined to form composite systems. For example, a hydrogen atom is a composite quantum system made by two quantum particles: a proton and an electron.

In quantum mechanics, composite systems are described in tensor product Hilbert spaces. Let S_A and S_B be quantum systems described in H_A and H_B respectively. The composite system $S_A + S_B$ made by S_A and S_B is described in the Hilbert space:

$$\mathsf{H} = \mathsf{H}_A \otimes \mathsf{H}_B.$$

Given a quantum state $\rho \in \mathfrak{B}(\mathsf{H})$, we can calculate the partial trace of ρ with respect to H_B (H_A) to obtain the corresponding *reduced state* of the quantum system S_A (S_B). In fact, given an observable A that can be measured on the system S_A, it can be identified to the observable $A \otimes \mathbb{I}_{\mathsf{H}_B}$ for the composite system. Its expectation value, on the state ρ of the composite system, is:

$$\langle A \rangle_{\rho, S_A} = \mathrm{tr}(\rho(A \otimes \mathbb{I}_{\mathsf{H}_B})) = \mathrm{tr}(\mathrm{tr}_{\mathsf{H}_B}(\rho)A), \qquad (2.5.1)$$

thus the partial trace $\mathrm{tr}_{\mathsf{H}_B}(\rho)$ is the density matrix encoding the outcome statistic of any observable that can be measured on S_A.

Let $|\psi\rangle \in \mathsf{H}_A$ and $|\varphi\rangle \in \mathsf{H}_B$ be pure states. Then $|\psi\rangle \otimes |\varphi\rangle \in \mathsf{H}_A \otimes \mathsf{H}_B$ is a state of the composite system $S_A + S_B$. However, an arbitrary pure state $|\Psi\rangle \in \mathsf{H}_A \otimes \mathsf{H}_B$ may be not a tensor product of two vectors but, in general, it is a linear combination of tensor products. If $|\Psi\rangle \in \mathsf{H}_A \otimes \mathsf{H}_B$ is in the product form:

$$|\Psi\rangle = |\psi\rangle \otimes |\varphi\rangle \qquad |\psi\rangle \in \mathsf{H}_A, |\varphi\rangle \in \mathsf{H}_B, \qquad (2.5.2)$$

then it is called **separable**, otherwise it is called **entangled**. If the pure state of $S_A + S_B$ is separable then the subsystems are uncorrelated and each of them presents a well-defined state. On the other hand if the state is entangled the system are correlated within a quantum superposition. We can give the more general definition of entanglement in terms of density operators.

Definition 2.5.1 *The density operator $\rho \in \mathfrak{S}(\mathsf{H}_A \otimes \mathsf{H}_B)$ is said to be **separable** if it can be written as a statistical mixture of product states:*

$$\rho = \sum_i \lambda_i \rho_i^{(A)} \otimes \rho_i^{(B)}, \qquad (2.5.3)$$

*where $\lambda_i \geq 0$ and $\sum_i \lambda_i = 1$, $\rho_i^{(A)} \in \mathfrak{S}(\mathsf{H}_A)$ and $\rho_i^{(B)} \in \mathfrak{S}(\mathsf{H}_B)$ $\forall i$. Otherwise it is said to be **entangled**.*

If ρ is a pure state then condition (2.5.3) reduces to (2.5.2) that is the definition of separability for pure states. Roughly speaking, if the quantum state of $S_A + S_B$ has form (2.5.3) then the states of the single subsystems S_A and S_B present a classical correlation due to the statistical mixture. If the state is entangled then S_A and S_B present a higher degree of correlation that is beyond the classical description of the statistical mixture (2.5.3).

In order to clarify that entangled states carry a kind of non-classical correlation, let us consider the following example of an entangled pure state of a qubit pair:

$$|\Psi\rangle = \frac{1}{\sqrt{2}}(|00\rangle + |11\rangle) \in \mathbb{C}^2 \otimes \mathbb{C}^2, \qquad (2.5.4)$$

where we have adopted the compact notation $|00\rangle \equiv |0\rangle \otimes |0\rangle$ and $|11\rangle = |1\rangle \otimes |1\rangle$. Let us assume to perform a measurement process on the first qubit, in particular we measure the observable described by the Pauli matrix σ_z. In other words we perform the measurement $\mathcal{M}$ described by the PVM $\{|0\rangle\langle 0|, |1\rangle\langle 1|\}$ associated to the computational basis. According to the notion of post-measurement state defined in (2.3.14), the state of the 2-qubit system after the measurement is:

$$|\Psi\rangle' = \begin{cases} |00\rangle & \text{if the outcome of } \mathcal{M} \text{ is } 0 \\ |11\rangle & \text{if the outcome of } \mathcal{M} \text{ is } 1. \end{cases}$$

Therefore, a local measurement on the first qubit affects the state of the other qubit regardless their spatial separation. Such a perturbation is instantaneous and violates the Einstein's principle of locality, this is the so-called EPR paradox [EPR35] proposed to show that quantum mechanics is not a complete theory. Unfortunately a discussion about extremely interesting topics like the EPR paradox, Bell's theorem, extension of quantum mechanics via hidden-variable theories, local realism, non-contextuality are beyond the scope of this volume. Nowadays the most accepted interpretation of quantum mechanics (mainly based on Bell's inequalities [Be64, Be66] and their experimental violation [ADR82, We98, He15]) contemplates the non-locality as an intrinsic characteristic of quantum phenomena. Thus entangled states enable non-classical effects that can be exploited in quantum information and quantum computing to define efficient methods of information transfer (quantum teleportation and superdense coding) and processing (quantum algorithms) that outperform classical counterparts. However, entangled states cannot be used for superluminal communications (*no-communication theorem* [Pe04]), so in this "no-signaling form" the Einstein's principle of locality is preserved and not violated by quantum correlations.

In section 2.4, we have seen that the time evolution of an isolated quantum system is *unitary*, since a unitary operator is invertible the quantum evolution is informationally lossless. However, this is true only when the evolving quantum system does not interact with the environment. In the case we have an open (that is non-isolated) system we must consider the composite system described in $\mathsf{H}_S \otimes \mathsf{H}_E$ where H_S is the Hilbert space of the considered quantum system S and H_E is the Hilbert space of another system that models the environment. Let ρ_0 and ρ_E be the initial state of S and environment respectively. Since the total system $S+environment$ is isolated, its evolution is described by a group of unitary operators $\{U_t\}_{t\in\mathbb{R}}$ and the reduced time evolution of the system S is obtained tracing out the environment:

$$\rho_t = \mathrm{tr}_{\mathsf{H}_E}[U_t(\rho_0 \otimes \rho_E)U_t^\dagger] \quad \text{for} \quad t > 0, \tag{2.5.5}$$

that is a non-unitary evolution allowing decoherence and information losses. As discussed in the next chapter, in quantum computing we describe operations on qubits by means of unitary operators, called *quantum gates*, under the assumption that the considered quantum hardware is noiseless and protected against decoherence.

Chapter 3

Basics of Quantum Computing

Quantum computing is a type of computation where quantum phenomena, such as state superposition and entanglement, are exploited to perform calculations. It is the most prominent application of quantum information theory and delivers algorithms to solve efficiently some problems which are hard for classical computers. This chapter is focused on the fundamentals of quantum computing like the abstract notion of a universal quantum computer and the circuit model for quantum computations. There is an overview on adiabatic quantum computing which provides a notion of analog quantum computer. Quantum annealers, as specific-purpose quantum machines, are also introduced.

3.1 Encoding data into quantum states

In quantum information theory, a *quantum encoding* is any procedure to map classical data into the physical states of a considered quantum system. In this section we list the quantum encodings that are relevant in the following to encode the values of discrete and continuous variables into quantum states.

The *basis encoding* is defined by mapping a collection of items into the states forming an orthonormal basis of the Hilbert space of the considered quantum system. Given a finite set X with cardinality $|X|$, any $x \in X$ can be encoded into a basis vector $|x\rangle$ of a Hilbert space of dimension $|X|$. Physically, the orthonormal basis $\{|x\rangle\}_{x \in X}$, called *computational basis*, is made by the eigenstates of a reference observable measured on the considered quantum system. For instance, a bit can be encoded into a qubit by the mapping $0 \mapsto |0\rangle$, $1 \mapsto |1\rangle$. Then the n-bit strings $(x_1 \cdots x_n)$ can be encoded into the states of n qubits forming an orthonormal basis of a 2^n-dimensional Hilbert space H_n:

$$\mathbb{B}^n \ni (x_1 \cdots x_n) \mapsto |x_1 \cdots x_n\rangle \in \mathsf{H}_n. \tag{3.1.1}$$

D. Pastorello, *Concise Guide to Quantum Machine Learning*,
Machine Learning: Foundations, Methodologies, and Applications,
https://doi.org/10.1007/978-981-19-6897-6_3

Then the qubits can be prepared in a superposition of different data that can be processed in parallel by linearity. In particular one can consider the superposition of all the possible binary strings of length n:

$$|\psi\rangle = \frac{1}{\sqrt{2^n}} \sum_{x=0}^{2^n-1} |x\rangle. \tag{3.1.2}$$

The initialization of the complete superposition (3.1.2) is the idea behind the quantum advantage of the celebrated Grover's search for instance.

The *amplitude encoding* is the representation of classical data into the amplitudes of a quantum state. A complex vector $\mathbf{x} \in \mathbb{C}^d$ with unit norm can be represented by the amplitudes of a quantum state $|\psi_{\mathbf{x}}\rangle$ with respect to a fixed basis $\{|\phi_i\rangle\}_i$ of the d-dimensional Hilbert space H:

$$|\psi_{\mathbf{x}}\rangle = \sum_{i=1}^{d} x_i|\phi_i\rangle \in \mathsf{H}. \tag{3.1.3}$$

If the classical information is given by a $d{\times}d$ complex matrix A satisfying $\sum_{ij} |a_{ij}|^2 = 1$, it can be encoded in the state:

$$|\psi_A\rangle = \sum_{i,j=1}^{d} a_{ij}|\phi_i\rangle \otimes |\phi_j\rangle \in \mathsf{H} \otimes \mathsf{H}. \tag{3.1.4}$$

The main advantage of the amplitude encoding is the space efficiency with respect to the basis encoding. Consider a system made by n qubits, within the basis encoding we can convey only n classical bits, within the amplitude encoding we can store complex vectors of dimension 2^n. The limitations of amplitude encoding are the normalization requirement on the classical data and the fact that quantum amplitudes x_i cannot be directly observed and only the real numbers $|x_i|^2$ can be retrieved from the state $|\psi_{\mathbf{x}}\rangle$ where the complex vector $\mathbf{x}$ is stored. Therefore, the amplitude encoding requires normalized data vectors or norms that are given separately.

A third kind of quantum encoding is related to probability distributions. Given a probability distribution p on the finite set X, it can be encoded in the state:

$$|\psi_p\rangle = \sum_{x\in X} \sqrt{p(x)}|x\rangle \in \mathsf{H}, \tag{3.1.5}$$

where $\{|x\rangle\}_{x\in X}$ is an orthonormal basis of the $|X|$-dimensional Hilbert space H. Repeated measurements on the state $|\psi_p\rangle$ with respect to the computational basis allow to sample the distribution p. Some authors call this procedure *qsample encoding*.

The properties of the quantum encoding are fundamental aspects of quantum computing, quantum communication schemes, quantum cryptography protocols. One of the most remarkable characteristic of quantum encoding is that data cannot be copied because unknown quantum states cannot be cloned.

Definition 3.1.1 *A **quantum cloner** (or **quantum cloning machine**) is a composite quantum system described in the Hilbert space* $\mathsf{H} \otimes \mathsf{H}$ *such that there is a state* $|\eta\rangle \in \mathsf{H}$ *and a unitary operator* U *satisfying:*

$$U|\psi\eta\rangle = |\psi\psi\rangle \quad \forall |\psi\rangle \in \mathsf{H}. \tag{3.1.6}$$

A quantum cloner is characterized by a time evolution that duplicates the state of a subsystem into the state of the other subsystem prepared in the "blank state" $|\eta\rangle$. We have the following no-go theorem [WZ82].

Theorem 3.1.2 (No-cloning) *A quantum cloner does not exist.*

Proof. Let us assume the existence of a quantum cloner and consider two arbitrary pure state $|\psi\rangle, |\varphi\rangle \in \mathsf{H}$. Then:

$$U|\psi\eta\rangle = |\psi\psi\rangle \quad \text{and} \quad U|\psi\eta\rangle = |\psi\psi\rangle. \tag{3.1.7}$$

Taking the inner products of the lefthand sides and the righthand sides of the equations (3.1.7), we have:

$$\langle\psi\eta|U^\dagger U|\varphi\eta\rangle = \langle\psi\eta|\varphi\eta\rangle = \langle\psi|\varphi\rangle\langle\eta|\eta\rangle = \langle\psi|\varphi\rangle \tag{3.1.8}$$

$$\langle\psi\psi|\varphi\varphi\rangle = \langle\psi|\varphi\rangle\langle\psi|\varphi\rangle = \langle\psi|\varphi\rangle^2. \tag{3.1.9}$$

The equality $\langle\psi|\varphi\rangle = \langle\psi|\varphi\rangle^2$ is true only for coinciding or orthogonal states therefore a unitary operator U satisfying (3.1.6) does not exist. $\qquad \square$

There are more general versions of Theorem 3.1.2 in the formalism of *quantum channels* [Li99, Pa19], however let us present only the elementary formulation above. The no-cloning theorem has several important consequences and it is one of the fundamental aspects of the security in quantum communications. Roughly speaking, without the possibility of copying intercepted quantum states an eavesdropper must extract information only performing measurements corrupting the transmission and revealing his/her presence. Quantum no-cloning is a resource for data security, however it is also a limitation because it prevents the existence of repeaters for long-range quantum communications. In quantum computing, it prevents backup copies of a quantum state for error correction during a computation. In quantum machine learning, if we consider the quantum implementation of a perceptron, the no-cloning limits the direct construction of a feed-forward neural network because the quantum output of a neuron cannot be copied to feed other neurons. The no-cloning theorem has a very general consequence that can be considered the statement of another no-go theorem: *a lossless conversion of quantum information into classical information is not allowed.* In fact, if there is a general prescription of performing a measurement on an arbitrary quantum state and using the outcomes to re-prepare exactly the initial quantum state then such a prescription is not consistent with no-cloning theorem.

Encoding data into quantum states enables several applications of the quantum phenomena to computation. In particular the initialization of data superpositions allows efficient computations by *quantum parallelism* and entanglement among quantum registers provides a kind of data processing without a classical counterpart. For example, the quantum implementation of the binary classifier proposed in [PB21] and described in Section 7.1 is based on the amplitude encoding of the training vectors that are put in superposition, also the unclassified vector is put in superposition of the two possible classes. Then the entire training set and the new feature vector are entangled with an ancillary qubit so that the execution of the classifier is done processing a single qubit regardless the size of the training set and dimension of the feature space.

3.2   Quantum circuits

The Church-Turing conjecture reads that any computable function can be computed by a *Turing machine*, then the Turing machine is said to be a universal model of classical computation [BL74]. There is a quantum analogue, the *quantum Turing machine*, that provides the notion of a universal quantum computer [De85]. Here we do not discuss the quantum Turing machine but we introduce an equivalent model, the *quantum circuit model*. The proof of the computational equivalence between quantum circuits and quantum Turing machine is presented in [Ya93].

Definition 3.2.1 *A system of n qubits, described in the Hilbert space $\mathsf{H}_n \simeq (\mathbb{C}^2)^{\otimes n}$, is called **$n$-qubit register** and any unitary operator on H_n is called **n-qubit gate**.*

From the physical viewpoint a quantum gate is a controlled time evolution of an isolated composite system made by qubits. Quantum gates admit a graphical representation like the classical logical gates then a quantum computation can be always represented by the analogue of a digital circuit and its time complexity can be evaluated in these terms. Let us list the main quantum gates, the *Hadamard gate* is a 1-qubit gate defined in matrix form with respect to the computational basis $\{|0\rangle, |1\rangle\}$ as follows:

$$H := \frac{1}{\sqrt{2}} \begin{pmatrix} 1 & 1 \\ 1 & -1 \end{pmatrix}, \tag{3.2.1}$$

its graphical representation is:

$$\begin{array}{c}\boxed{H}\end{array} \qquad .$$

The Hadamard gate realizes a change of basis $\{|0\rangle, |1\rangle\} \mapsto \{|+\rangle, |-\rangle\}$ of a 1-qubit Hilbert space where $|+\rangle$ and $|-\rangle$ are defined in (2.3.23).

The 1-qubit gate that appends a relative phase in the input state, it is defined by:

$$P_\phi := \begin{pmatrix} 1 & 0 \\ 0 & e^{i\phi} \end{pmatrix}, \tag{3.2.2}$$

where $\phi \in \mathbb{R}$, its graphical representation is:

$$\begin{array}{c}\boxed{P_\phi}\end{array} \qquad .$$

The 1-qubits gates defined by $S := P_{\pi/2}$ and $T := P_{\pi/4}$ play a crucial role in defining a universal set of quantum gates.

The following statement entails a characteristic decomposition of 1-qubit gates that is crucial for constructing controlled quantum gates [CN00].

Proposition 3.2.2 *For any 1-qubit gate U there exist unitary operators A, B, C satisfying $ABC = \mathbb{I}$ and $\alpha \in \mathbb{R}$ such that:*

$$U = e^{i\alpha} A\sigma_x B\sigma_x C. \tag{3.2.3}$$

The prototypical controlled operation is the controlled-NOT (CNOT). The CNOT gate is a 2-qubit gate defined, with respect to the computational basis, by:

$$\mathsf{CNOT}|x\rangle|y\rangle := |x\rangle|y \oplus x\rangle, \qquad x, y \in \{0, 1\}, \tag{3.2.4}$$

so the first qubit controls the conditional action of a bit-flip on the second qubit, its graphical representation is:

More generally, given a 1-qubit gate U, a controlled quantum gate CU can be defined by $CU|x\rangle|y\rangle := |x\rangle U^x|y\rangle$ and denoted by the symbol:

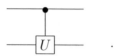

In view of the decomposition (3.2.3) of U, the controlled gate CU can be implemented as follows:

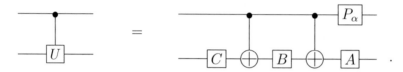

The SWAP gate is a 2-qubit gate that swaps the input qubits. It is obtained combining three CNOTs:

$$(3.2.5)$$

The controlled SWAP:

$$(3.2.6)$$

is called *Fredkin gate* as its classical version that is universal for classical reversible computing.

In quantum computations, measurement processes obviously play a crucial role because they enable the readout of the results and realize projection operations over the quantum registers. In quantum circuits, the symbol:

denotes a measurement process with respect to the computational basis. The action of a quantum gate can be controlled by the outcome of a measurement, in this case the corresponding circuit diagram is:

In classical digital computing, a small set of logical gates, like {AND, OR, NOT}, can be used to compute an arbitrary function. In quantum computing, there is a similar notion of universality that takes into account that the group of the unitary operators on a Hilbert space is continuous, so a finite set of gates cannot exactly implement any unitary operator.

Definition 3.2.3 *A set of quantum gates is said to be **universal for quantum computation** if, for all $n \in \mathbb{N}$, any n-qubit gate can be approximated to arbitrary accuracy by a composition of only those gates.*

A proof of the following theorem can be found in [CN00], the statement provides the so-called *standard* universal set for quantum computation.

Theorem 3.2.4 *The set $\{H, S, T, \mathsf{CNOT}\}$ is universal for quantum computation.*

Assuming that the gates of a universal set require fixed amounts of time to be implemented on the qubits, time complexity of a quantum algorithm can be estimated by the gate counting in the corresponding circuit. In complexity theory, the *time complexity* of an algorithm is the relation between the computation time t and the binary size L of the input data. The *classical* time complexity of a problem is defined as the infimum of the complexities of all the *classical* algorithms that can solve the problem. A precise definition of *computation time* requires a notion of elementary operation within a computational model like a single head movement in a Turing machine or the execution of a logical gate in a digital circuit. However, the strong Church-Turing hypothesis states that each model can be simulated in polynomial time by a probabilistic Turing machine then the definition of complexity classes is model-independent. The class **P** contains all the decision problems (that is, the solution is either *yes* or *no*) that can be solved in polynomial time in the sense that t, as a function of L, is bounded from above by a polynomial, using the big-O notation we write $t = O(L^a)$ with $a > 0$. When we say that a problem is "easy" and a computation is "efficient" we are referring to class **P**. A decision problem, described by the boolean function f, is in class **NP** if there is a decision problem, described by the boolean function g, in class **P** such that: $g(x, y) = 1$ for some $y \Rightarrow f(x) = 1$. Obviously **P**$\subseteq$ **NP** and the widely believed conjecture is that **P** $\neq$ **NP**. A decision problem P is in the class **NP-hard** if and only if any **NP** problem can be reduced to P in polynomial time. The class **NP-complete** contains all the decision problems in **NP** that are **NP-hard**. Let us consider a probabilistic computation described by the transition matrix $T = \{T_{xy}\}$ where $y \mapsto T_{xy}$ is the probability distribution of the output given the input x. Due to the probabilistic output, the computation may return a wrong result. The class **BPP** contains all the decision problems that can be solved in polynomial time with error probability less than $1/3$. We have that $\mathbf{P} \subseteq \mathbf{BPP}$ but the relation between $\mathbf{BPP}$ and $\mathbf{NP}$ is not know.

The complexity classes introduced so far are based on the notion of *classical computations* that can be considered a special case of *quantum computations*. Let us clarify this point: Let $f : \mathbb{B}^n \to \mathbb{B}^m$ be a computable function that is non-invertible, f can be always re-defined as an invertible function $\tilde{f} : \mathbb{B}^{n+m} \to \mathbb{B}^{n+m}$ defined by $\tilde{f}(x,y) := (x, y \oplus f(x))$. Thus an irreversible computation can be always simulated efficiently by a reversible computation for which the input can be reconstructed from the output. Since any invertible function $f : \mathbb{B}^n \to \mathbb{B}^n$ defines a unitary operator U_f on a n-dimensional Hilbert space by the following action with respect to the computational basis $U_f : |x\rangle \mapsto |f(x)\rangle$, we have that any classical computation can be efficiently performed by a quantum computer.

Let us define the quantum complexity class $\mathbf{BQP}$ of decision problems that can be solved by a quantum computer in polynomial time with error probability less than $1/3$. Since any classical computation can be simulated by a quantum computation we have $\mathbf{P} \subseteq \mathbf{BPP} \subseteq \mathbf{BQP}$. The conjecture $\mathbf{BPP} \neq \mathbf{BQP}$ is motivated, for instance, by the *Shor's algortihm* that solves the integer factoring problem (that is generally suspected to be not in $\mathbf{P}$) in polynomial time. Assuming $\mathbf{P} \neq \mathbf{NP}$, the exact relationship between $\mathbf{BQP}$ and $\mathbf{NP}$ is not known.

3.3   Quantum oracles

In computability theory, an *oracle* is defined as an abstract *black box* that computes a function $f : \mathbb{B}^n \to \mathbb{B}^m$ as a single operation. Its graphical representation is:

$$ a \;\; \rule{1cm}{0.4pt}\boxed{f}\rule{1cm}{0.4pt}\;\; f(a) \;\; . \tag{3.3.1} $$

Since f is not necessarily invertible, one can re-define the oracle as in (3.3.2) in order to perform a *reversible computation*. As mentioned in the previous section, we are particularly interested in reversible computing because quantum gates (that are unitary transformations) are always reversible. The invertible version of (3.3.1) presents an input register and an output register:

$$ \begin{array}{c} a \\ \\ b \end{array} \quad \boxed{f} \quad \begin{array}{c} a \\ \\ b \oplus f(a), \end{array} \tag{3.3.2} $$

where $\oplus$ is the sum modulo 2. Let us consider the basis encoding of the binary variables a, b and $b \oplus f(a)$ into $|a\rangle \in \mathsf{H}_{in}$, $|b\rangle$ and $|b \oplus f(a)\rangle \in \mathsf{H}_{out}$. As mentioned in the previous section, any invertible function defines a unitary operator, so the *quantum oracle* is an operator on $\mathsf{H}_{in} \otimes \mathsf{H}_{out}$ defined as:

$$U_f(|a\rangle \otimes |b\rangle) := |a\rangle \otimes |b \oplus f(a)\rangle. \tag{3.3.3}$$

If H_{in} is a n-qubit register and H_{out} is a m-qubit register, the quantum oracle is a $(n + m)$-qubit gate:

$$
\begin{array}{c}
|a\rangle \quad\rule{0pt}{0pt}\quad U_f \quad\rule{0pt}{0pt}\quad |a\rangle \\
|b\rangle \quad\rule{0pt}{0pt}\quad\quad |b \oplus f(a)\rangle \ .
\end{array}
\tag{3.3.4}
$$

Let us give a simple example to illustrate a quantum oracle at work: Consider the following decision problem: *Is the function $f : \{0,1\} \to \{0,1\}$ constant?* In order to solve this simple problem one needs two calls to the oracle, that is one must give both possible inputs in order to check if $f(0) = f(1)$ or $f(0) \neq f(1)$. This problem can be submitted to a *quantum oracle* implementing f that is the 2-qubit gate defined as:

$$
\begin{array}{c}
|x\rangle \quad\rule{0pt}{0pt}\quad U_f \quad\rule{0pt}{0pt}\quad |x\rangle \\
|y\rangle \quad\rule{0pt}{0pt}\quad\quad |y \oplus f(x)\rangle
\end{array}
\tag{3.3.5}
$$

where $x, y \in \{0,1\}$. By the submission of the input state $|+\rangle \otimes |-\rangle$ we have:

$$
\begin{array}{c}
|+\rangle \quad\rule{0pt}{0pt}\quad U_f \quad\rule{0pt}{0pt}\quad \frac{(-1)^{f(0)}|0\rangle + (-1)^{f(1)}|1\rangle}{\sqrt{2}} \\
|-\rangle \quad\rule{0pt}{0pt}\quad\quad |-\rangle
\end{array}
\tag{3.3.6}
$$

Therefore, the state of the first qubit in output is $|+\rangle$ if $f(0) = f(1)$ or $|-\rangle$ if $f(0) \neq f(1)$ up to the sign (that is a non-observable global phase). A single measurement on the first qubit is sufficient to decide if f is constant, then the decision problem is solved by only one query.

Generalizing the discussed example we can introduce the *Deutsch-Jozsa algorithm* [DJ92, Cl98]. Let us consider a function $f : \{0,1\}^n \to \{0,1\}$ that is either constant or balanced (that is, it attains value 0 on 2^{n-1} inputs). Now we consider the problem to decide if f is constant or balanced. If f is implemented by a classical oracle, in the worst case, the number of oracle calls that is required to solve the problem is $2^{n-1} + 1$. If a quantum oracle is available, the following circuit can be executed over a $(n + 1)$-qubit register:

$$|0\rangle^{\otimes n} \;-\boxed{H^{\otimes n}}\;-\boxed{\;\;U_f\;\;}\;-\boxed{H^{\otimes n}}\;-\boxed{\measuredangle} \qquad\qquad (3.3.7)$$

$$|1\rangle \;-\boxed{H}\;-$$

The action of the $n + 1$ Hadamard gates on the initial state is:

$$H^{\otimes(n+1)}|0\cdots01\rangle = \frac{1}{\sqrt{2^n}} \sum_{x=0}^{2^n-1} |x\rangle \otimes \frac{|0\rangle - |1\rangle}{\sqrt{2}},$$

the quantum oracle produces the state:

$$U_f\left(\frac{1}{\sqrt{2^n}} \sum_{x=0}^{2^n-1} |x\rangle \otimes \frac{|0\rangle - |1\rangle}{\sqrt{2}}\right) = \frac{1}{2^{\frac{2n}{2}}} \sum_{x=0}^{2^n-1} (-1)^{f(x)}|x\rangle \otimes \frac{|0\rangle - |1\rangle}{\sqrt{2}},$$

and the action of the Hadamard gate on the first n qubits is:

$$(H^{\otimes n}\otimes\mathbb{I})\left(\frac{1}{2^{\frac{2n}{2}}} \sum_{x=0}^{2^n-1} (-1)^{f(x)}|x\rangle \otimes \frac{|0\rangle - |1\rangle}{\sqrt{2}}\right) = \frac{1}{2^n}\left(\sum_{x=0}^{2^n-1}\sum_{y=0}^{2^n-1} (-1)^{f(x)+x\cdot y}|y\rangle\right)\otimes\frac{|0\rangle - |1\rangle}{\sqrt{2}}.$$

The probability that a measurement, in the computational basis, on the n-qubit register, returns the string of n zeros is:

$$\mathbb{P}(0) = \left| \frac{1}{2^n} \sum_{x=0}^{2^n-1} (-1)^{f(x)} \right|^2.$$

Therefore, $\mathbb{P}(0) = 1$ if f is constant and $\mathbb{P}(0) = 0$ if f is balanced. A single oracle call allows to solve the decision problem. There are not practical applications of this quantum algorithm, however it is a clear illustration of the impact of quantum oracles. More precisely, once defined the class **QP** of all the decision problems that can be deterministically solved in polynomial time by a quantum computer, we have $\mathbf{P} \neq \mathbf{QP}$.

3.4   Adiabatic quantum computing

Adiabatic quantum computing (AQC) was proposed as an application of *quantum adiabatic theorem* to solve optimization problems [FGG00], indeed it turns out to be equivalent to the quantum circuit model so it is a universal model for quantum computation [Ah07].

Let us briefly review the content of the adiabatic theorem: Assume to prepare a quantum system in the ground state $|\psi_0\rangle$ of a given Hamiltonian $\mathcal{H}$ (that is, $|\psi_0\rangle$ is the eigenstate with minimum eigenvalue $\mathcal{H}$), then one can change the Hamiltonian smoothly in time. If the change is sufficiently slow then the system remains in the istantaneous ground state with high probability. The time-dependent Hamiltonian is described by a smooth one-parameter family of self-adjoint operators $\{\mathcal{H}(t)\}_{t\in[0,T]}$ in the Hilbert space H of the considered quantum system. The dynamics of the system that is prepared in the ground state $|\psi_0\rangle \in \mathsf{H}$ is given by the solution of the Schrödinger equation:

$$i\hbar\frac{d}{dt}|\psi(t)\rangle = \mathcal{H}(t)|\psi(t)\rangle \qquad t \in [0,T], \tag{3.4.1}$$

with the initial condition $|\psi(0)\rangle = |\psi_0\rangle$. Let us re-parametrize the time-dependent Hamiltonian as $\widetilde{\mathcal{H}}(s) := \mathcal{H}(Ts)$ with $s \in [0,1]$. For any $s \in [0,1]$ we have the following eigenvalue problem:

$$\widetilde{\mathcal{H}}(s)|l,s\rangle = E_l(s)|l,s\rangle, \tag{3.4.2}$$

where $E_0(s)$ is the minimum of the spectrum of $\widetilde{\mathcal{H}}(s)$ and $|0,s\rangle$ the corresponding eigenvector, that is the ground state. Let us assume the non-degeneracy of the initial ground state, $\dim \mathsf{H}_{E_0(0)} = 1$, where $\mathsf{H}_{E_0(0)}$ is the eigenspace of $E_0(0)$. A simple formulation of the adiabatic theorem, assuming the non-degeneracy of the ground state for any $s \in [0,1]$, is the following [Ka50, BF28]:

Theorem 3.4.1 *If $\lambda(s) := E_1(s) - E_0(s) > 0$ for any $s \in [0,1]$ then:*

$$\lim_{T\to+\infty} |\langle 0,1|\psi(T)\rangle| = 1, \tag{3.4.3}$$

where $\psi(T)$ is the solution of (3.4.1), with initial condition $\psi(0) = |0,0\rangle$, calculated in $t = T$.

The existence of a nonzero spectral gap λ ensures that the state of the evolving system remains in the ground state of $\mathcal{H}(t)$, for any $t \in [0,T]$, if the evolution time is large.

In order to evaluate how big T would be to obtain an acceptable probability to remain in the ground state, one can use the following rough estimation [Mc14] but more refined adiabatic conditions exist [JRS07]:

$$T \gg \frac{\max_s \| \frac{d}{ds}\widetilde{\mathcal{H}}(s) \|_{op}}{[\min_s \lambda(s)]^2}, \tag{3.4.4}$$

where the operator norm $\| \cdot \|_{op}$ is defined in (2.2.3).

The general structure of an adiabatic quantum computation is the following:

- A quantum system is prepared in the known ground state of an initial Hamiltonian $\mathcal{H}_I$ such that $[\mathcal{H}_I, \mathcal{H}_P] \neq 0$.

- A time variation of the Hamiltonian from $\mathcal{H}_I$ to $\mathcal{H}_P$ is implemented according to:

$$\mathcal{H}(t) = (1 - s(t))\mathcal{H}_I + s(t)\mathcal{H}_P \quad t \in [0, T] \tag{3.4.5}$$

  where T satisfies (3.4.4), $s : [0, \tau] \to [0, 1]$ is a smooth monotone function such that $s(0) = 0$ and $s(T) = 1$ and $\mathcal{H}_P$ is the problem Hamiltonian.

- Measurement process on the quantum system.

The problem to solve is encoded in $\mathcal{H}_P$ and its ground state represents the solution. By the adiabatic theorem, the final measurement process is performed on the ground state of $\mathcal{H}_P$ with high probability. In order to formulate an adiabatic algorithm one must to encode the problem into $\mathcal{H}_P$ and estimate the spectral gap λ to perform the adiabatic evolution that cannot be too slow in order to not destroy the efficiency of the algorithm.

3.5 Quantum annealing

Quantum annealing (QA) is a type of heuristic search used to solve optimization problems [KN98]. The solution of a given problem corresponds to the *ground state* of a quantum system with total energy described by a *problem Hamiltonian* $\mathcal{H}_P$ on the Hilbert space where the considered quantum system is described. The annealing procedure is implemented by a time evolution of the quantum system towards the ground state of the problem Hamiltonian. QA is related to AQC by some crucial aspects but the two techniques do not coincide. In AQC the considered quantum system is assumed to be isolated so its dynamics is unitary, in QA the quantum hardware is considered as an open system interacting with the environment then its dynamics is characterized by decoherence and energy dissipation.

Let us consider the time-dependent Hamiltonian

$$\mathcal{H}(t) = \Gamma(t)\mathcal{H}_D + \mathcal{H}_P, \tag{3.5.1}$$

where $\mathcal{H}_P$ is the problem Hamiltonian, and $\mathcal{H}_D$ is the *transverse field Hamiltonian* (or disordering Hamiltonian), which gives the kinetic term inducing the exploration of the solution landscape by means of quantum fluctuations. Γ is a decreasing function that attenuates the kinetic term driving the system towards the global minimum of the problem landscape represented by $\mathcal{H}_P$. As in the D-Wave machines [DW21a], QA can be physically realized considering a quantum spin glass that is a network of qubits arranged on the vertices of a graph (V, E), with $|V| = n$, whose edges E represent the couplings among the qubits. The problem Hamiltonian is defined by:

$$\mathcal{H}_P = \sum_{i \in V} \theta_i \sigma_z^{(i)} + \sum_{(i,j) \in E} \theta_{ij} \sigma_z^{(i)} \sigma_z^{(j)}, \tag{3.5.2}$$

where the coefficient θ_i physically corresponds to the local field on the ith qubit and θ_{ij} to the coupling between the qubits i and j. The Hamiltonian (3.5.2) is an operator on the n-qubit Hilbert space $\mathsf{H} = (\mathbb{C}^2)^{\otimes n}$ where $\sigma_z^{(i)}$ acts as the Pauli matrix:

$$\sigma_z = \begin{pmatrix} 1 & 0 \\ 0 & -1 \end{pmatrix} \tag{3.5.3}$$

on the ith tensor factor and as the 2×2 identity matrix on the other tensor factors. Since the Pauli matrix σ_z has two eigenvalues $\{-1, 1\}$, the Hamiltonian (3.5.2) has the spectrum corresponding to the values of the cost function given by the energy of the well-known *Ising model*:

$$\mathsf{E}(\boldsymbol{z}) = \sum_{i \in V} \theta_i z_i + \sum_{(i,j) \in E} \theta_{ij} z_i z_j, \quad \boldsymbol{z} = (z_1, ..., z_n) \in \{-1, 1\}^{|V|}. \tag{3.5.4}$$

In practice, the annealing procedure drives the system into the ground state of the Hamiltonian $\mathcal{H}_P$ which corresponds to the solution of the considered problem:

$$\boldsymbol{z}^* = \operatorname*{argmin}_{\boldsymbol{z} \in \{-1,1\}^{|V|}} \mathsf{E}(\boldsymbol{z}). \tag{3.5.5}$$

Given a problem, the annealer is initialized by a suitable choice of the weights θ_i and θ_{ij}, the variables $z_i \in \{-1, 1\}$ are physically realized by the outcomes of the measurements on the qubits located in the vertices of the graph representing the hardware architecture. In order to solve a given optimization problem by QA, one needs to obtain the correct *encoding* of the objective function in terms of the cost function (3.5.4), which is hard in general. Detailed discussions about QA, its realization by means of spin glasses and its application to solve optimization problems can be found in [DC05, KN98, MN08] for instance.

The optimization problems that are usually considered in quantum annealing are the Quadratic Unconstrained Binary Optimization (QUBO) problems that are NP-hard problems traditionally used in computer science. A general QUBO problem is defined by the minimization of a function $f : \{0,1\}^n \to \mathbb{R}$ of the form:

$$f(x) = \sum_{i=1}^{n} Q_{ii}x_i + \sum_{i<j} Q_{ij}x_i x_j, \qquad (3.5.6)$$

where Q_i and Q_{ij} are real coefficients that can be arranged into an $n \times n$ upper-triangular real matrix Q. Therefore, a QUBO problem can be represented as:

$$\min_{x \in \{0,1\}^n} x^T Q x, \qquad (3.5.7)$$

and cast into the annealer architecture. The direct representation of the problem into the quantum hardware is not possible in general because of the low connected topology of the qubit network. There are embedding techniques like the *minor embedding* by D-Wave [DW21b, DW21c] or the *quantum annealing learning search* [PB19] devised to represent a general QUBO problem into the annealer architecture. The QUBO model (3.5.7) covers a remarkable range of applications in combinatorial optimization: optimization problems on graphs, facility location problems, resource allocation problems, clustering problems, set partitioning problems, various forms of assignment problems, sequencing and ordering problems [Ko14].

Chapter 4

Relevant quantum algorithms

We are in position to give a short but satisfactory overview on some well known quantum algorithms that are relevant as subroutines in machine learning schemes. In particular, the quantum Fourier transform is a quantum implementation of the discrete Fourier transform [Co94], Grover's algorithm and amplitude amplification are quantum search algorithms in an unsorted database [Gr96, BH97], the phase estimation algorithm allows to estimate the eigenvalue corresponding to an eigenvector of a unitary operator [CN00] with remarkable quantum machine learning applications.

4.1 Quantum Fourier transform

The *discrete Fourier transform* (DFT) is a map $\mathcal{F} : \mathbb{C}^N \to \mathbb{C}^N$ defined by:

$$\mathcal{F} : (\alpha_0, ..., \alpha_{N-1}) \longmapsto (\beta_0, ..., \beta_{N-1}) \qquad \text{with} \qquad \beta_y := \frac{1}{\sqrt{N}} \sum_{x=0}^{N-1} \alpha_x e^{i\left(\frac{2\pi}{N} x \cdot y\right)}. \quad (4.1.1)$$

The direct calculation of DFT requires $O(N^2)$ operations, however the complexity of the classical *fast Fourier transform algorithm* is $O(N \log N)$.

Adopting the amplitude encoding of normalized complex vectors into the states of n qubits, with $N = 2^n$, the *quantum Fourier transform* (QFT) is defined as a unitary operator acting on a n-qubit register as follows:

$$U_{\mathcal{F}} \left[\sum_{x=0}^{2^n-1} \alpha_x |x\rangle \right] = \sum_{y=0}^{2^n-1} \beta_y |y\rangle \qquad \beta_y := \frac{1}{\sqrt{2^n}} \sum_{x=0}^{2^n-1} \alpha_x e^{i\left(\frac{2\pi}{2^n} x \cdot y\right)}. \quad (4.1.2)$$

Since the quantum amplitudes cannot be observed, the output of QFT cannot be directly read but it can be used for a further processing. In this sense QFT must be

D. Pastorello, *Concise Guide to Quantum Machine Learning*,
Machine Learning: Foundations, Methodologies, and Applications,
https://doi.org/10.1007/978-981-19-6897-6_4

intended as a part of a larger quantum algorithm like *Shor* [Sh94] or the *quantum phase estimation* (Section 4.4). The action of $U_{\mathcal{F}}$ on an element of the computational basis is:

$$U_{\mathcal{F}}|x\rangle = \frac{1}{\sqrt{2^n}} \sum_{y=0}^{2^n-1} e^{i\frac{2\pi}{2^n}x \cdot y}|y\rangle. \tag{4.1.3}$$

Let us write (4.1.3) making explicit the binary representation $y_1 \cdots y_n$ of the integer y:

$$U_{\mathcal{F}}|x\rangle = \frac{1}{\sqrt{2^n}} \sum_{y_1=0}^{1} \cdots \sum_{y_n=0}^{1} \exp\left[i\frac{2\pi}{2^n}x \cdot \sum_{j=1}^{n} 2^{n-j}y_j\right] |y_1 \cdots y_n\rangle \tag{4.1.4}$$

$$= \frac{1}{\sqrt{2^n}} \sum_{y_1=0,1} \cdots \sum_{y_n=0,1} \bigotimes_{j=1}^{n} \exp\left[i\frac{2\pi}{2^j}x \cdot y_j\right] |y_j\rangle = \bigotimes_{j=1}^{n} \frac{|0\rangle + e^{i\frac{2\pi}{2^j}x}|1\rangle}{\sqrt{2}},$$

Now let us consider the following circuit with four qubits, the generalization to n qubits is straightforward:

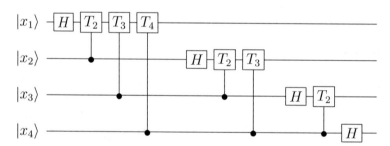

where $T_k := P_{\frac{2\pi}{2^k}}$ for $k = 1, 2, 3, 4$. The evolution of the first qubit is:

$$T_4T_3T_2H|x_1\rangle = T_4T_3T_2\frac{|0\rangle + (-1)^{x_1}|1\rangle}{\sqrt{2}} = T_4T_3\frac{|0\rangle + e^{i\pi x_1 + i\frac{\pi}{2}x_2}|1\rangle}{\sqrt{2}} =$$

$$= \frac{|0\rangle + e^{i2\pi \sum_{j=1}^{4}\frac{x_j}{2^j}}|1\rangle}{\sqrt{2}} = \frac{|0\rangle + e^{i\frac{2\pi x}{2^4}}|1\rangle}{\sqrt{2}},$$

where x is the integer with binary representation $x_1x_2x_3x_4$. The other three qubits evolves to the following states:

$$T_3T_2H|x_2\rangle = \frac{|0\rangle + e^{i\frac{2\pi x}{2^3}}|1\rangle}{\sqrt{2}},$$

$$T_2 H |x_3\rangle = \frac{|0\rangle + e^{i\frac{2\pi x}{2^2}} |1\rangle}{\sqrt{2}},$$

$$H |x_4\rangle = \frac{|0\rangle + e^{i\frac{2\pi x}{2}} |1\rangle}{\sqrt{2}}.$$

Therefore, the action of $U_{\mathcal{F}}$, for $n = 4$, is realized by the following circuit:

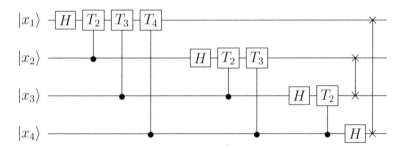

In the general case of n qubits, the ith qubit is processed by $n - i$ controlled phase gates and one Hadamard gate then the counting of the 1-qubit gates is

$$n + \sum_{i=1}^{n} (n - i) \sim n^2, \tag{4.1.5}$$

moreover there are $[n/2]$ SWAP gates, corresponding to $3[n/2]$ CNOT gates. Thus the complexity of QFT is $O(n^2)$. While the classical FFT algorithm calculates the discrete Fourier transform in time $O(n2^n)$ then there is an exponential speed-up.

4.2 Grover's search algorithm

The Grover's algorithm performs a search of a target item into an unsorted database of N objects in time $O(\sqrt{N})$, that is, with a quadratic speed-up with respect to a classical exhaustive search [Gr96].

Let $\{y_i\}_{i=0,\dots,N-1}$ be a collection of N items and assume to have an oracle $f : \{0, \dots, N-1\} \to \{0,1\}$ that recognize the solution: $f(x) = 1 \Leftrightarrow y_x$ is the target item. To find the target classically we must call the oracle $O(N)$ times. Using the basis encoding, the items $\{y_i\}_{i=0,\dots,N-1}$ can be encoded into the orthogonal quantum states $\{|x\rangle\}_{x=0,\dots,N-1} \subset \mathsf{H}_n$ of a n-qubit register with $n = \lceil \log N \rceil$. We can define a unitary operator that recognizes the target item as:

$$U_f |x\rangle := (-1)^{f(x)} |x\rangle. \tag{4.2.1}$$

The gate U_f appends a phase π (phase flip) to the state that encodes the label of the target. We define the *Grover's iteration* G as the operator:

$$G := H^{\otimes n} M H^{\otimes n} U_f \qquad \text{with} \qquad M := 2|0\rangle\langle 0| - \mathbb{I}_n, \qquad (4.2.2)$$

where $\mathbb{I}_n$ is the identity operator on H_n. The operator M acts as a phase flip to any state of the computational basis except $|0\rangle$. The operator U_f is realized by the quantum oracle (that is a $(n+1)$-qubit gate) as following:

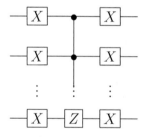

In fact the action of this gate is:

$$Qf(|x\rangle \otimes |-\rangle) = \frac{1}{\sqrt{2}}(|x\rangle \otimes |f(x)\rangle - |x\rangle \otimes |1 + f(x)\rangle)$$

$$= |x\rangle \otimes \frac{1}{\sqrt{2}}(|f(x)\rangle - |1 + f(x)\rangle) = (-1)^{f(x)}|x\rangle \otimes |-\rangle = U_f|x\rangle \otimes |-\rangle.$$

The action of M is implemented by the n-qubit gate:

$$
\begin{array}{ccccc}
\fbox{X} & \bullet & \fbox{X} \\
\fbox{X} & \bullet & \fbox{X} \\
\vdots & \vdots & \vdots \\
\fbox{X} & \fbox{Z} & \fbox{X}
\end{array}
$$

where the gates X and Z correspond to the application of Pauli matrices σ_x and σ_z respectively.

The Grover's algorithm is based on the repeated applications of G to the balanced state $|\Psi\rangle = \frac{1}{\sqrt{N}} \sum_x |x\rangle$ moving the state towards the solution $|\hat{x}\rangle$. Now let us calculate how many Grover's iterations are required to find the target with high probability following a simple geometric argument. The initial state can be written as:

$$|\Psi\rangle = \sin(\theta)|\hat{x}\rangle + \cos(\theta)|\varphi\rangle, \qquad (4.2.3)$$

where:

$$\theta = \arcsin\left(\frac{1}{\sqrt{N}}\right) \quad \text{and} \quad |\varphi\rangle := \frac{1}{\sqrt{N-1}} \sum_{x \neq \hat{x}} |x\rangle. \qquad (4.2.4)$$

By definition of U_f and M, we have:

$$U_f = \mathbb{I}_n - 2|\hat{x}\rangle\langle\hat{x}| \qquad \text{and} \qquad H^{\otimes n}MH^{\otimes n} = 2|\Psi\rangle\langle\Psi| - \mathbb{I}_n. \qquad (4.2.5)$$

Thus U_f is a reflection across $|\varphi\rangle$ in the hyperplane span$\{|\hat{x}\rangle, |\varphi\rangle\}$, since θ is the angle between $|\Psi\rangle$ and $|\varphi\rangle$ then 2θ is the angle between $|\Psi\rangle$ and $U_f|\Psi\rangle$. Then $H^{\otimes n}MH^{\otimes n}$ is the reflection across $|\Psi\rangle$ in span $\{|\hat{x}\rangle, |\varphi\rangle\}$. Therefore, G rotates $|\Psi\rangle$ by an angle 2θ towards $|\hat{x}\rangle$. After k Grover's iterations the resulting state is:

$$|\Psi\rangle_k = \sin[(2k+1)\theta]|\hat{x}\rangle + \cos[(2k+1)\theta]|\varphi\rangle, \qquad (4.2.6)$$

performing a measurement with respect to the computational basis, the probability to find the solution is:

$$\mathbb{P}_k(\hat{x}) = \sin^2[(2k+1)\theta] \quad \text{with} \quad \theta = \arcsin\left(\frac{1}{\sqrt{N}}\right) \simeq \frac{1}{\sqrt{N}}, \qquad (4.2.7)$$

if $k = \frac{\pi}{4}\sqrt{N} - \frac{1}{2}$ then $\mathbb{P}_k(\hat{x}) = 1$. Therefore, the number of Grover iterations, that is the number of oracle calls, to find the solution with maximum probability is $k = \lfloor \frac{\pi}{4}\sqrt{N} \rfloor$. Thus $O(\sqrt{N})$ is the query complexity of the quantum search in an unstructured database.

Example 4.2.1 *Consider a database of $N = 4$ items encoded in a 2-qubit register. A single query is sufficient to find the solution with maximum probability (Fig. 4.1).*

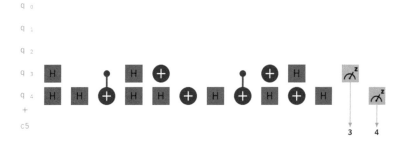

Figure 4.1. *The case $N = 4$ (2 qubits), the solution 11 is found with a single Grover iteration. The simple experiment can be done with the IBM Quantum Composer [IBM].*

Running the algorithm on the ibmq_16_melbourne quantum processor, a measurement on the 2-qubit register returns the target element with high probability (Fig. 4.2). The algorithm must be run multiple times in order to compute the statistic to find the most likely outcome.

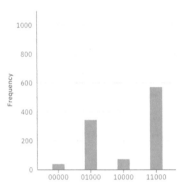

Figure 4.2. *Output statistic produced by the circuit in Fig. 4.1 over 1024 shots on the IBM quantum processor ibmq_16_melbourne.*

4.3 Amplitude amplification

The *amplitude amplification algorithm* is a generalization of the Grover's algorithm [BH97]. Given a database of N items and M target items to find, the algorithm returns a target element in time $O(\sqrt{N/M})$. All the items are put in quantum superposition then the initial state evolves towards a superposition of the target items, in other words the amplitudes of the targets are amplified and the other amplitudes are attenuated.

The items of the database X are encoded into the basis vectors $\{|x\rangle\}_{x \in X}$ of a $|X|$-dimensional Hilbert space H of a considered quantum system. Let $f : X \to \{0, 1\}$ be the oracular function such that $f(x) = 1$ if and only if x is a target element. We can define the *good subspace* H_g of H spanned by the target solutions:

$$\mathsf{H}_g := \mathrm{span}\{|x\rangle \in \mathsf{H} : x \in T\} \quad \text{where} \quad T = \{x \in X : f(x) = 1\}, \qquad (4.3.1)$$

and the *bad subspace* $\mathsf{H}_b := \mathsf{H}_g^\perp$. A state $|\psi\rangle \in \mathsf{H}$ can be written as:

$$|\psi\rangle = \sin\theta |\varphi_g\rangle + \cos\theta |\varphi_b\rangle \qquad \theta \in [0, \pi/2], \qquad (4.3.2)$$

where $|\varphi_g\rangle \in \mathsf{H}_g$ and $|\varphi_b\rangle \in \mathsf{H}_b$. Let us assume that $|\psi\rangle \notin \mathsf{H}_g, \mathsf{H}_b$, that is $\theta \neq 0, \pi/2$.

We can define the following unitary operators:

$$U_\psi := \mathbb{I} - 2|\psi\rangle\langle\psi| \quad , \quad U_g := \mathbb{I} - 2\sum_{x \in T} |x\rangle\langle x|, \qquad (4.3.3)$$

U_ψ acts as a phase flip on $|\psi\rangle$ and U_g acts as a phase flip on H_g. Let us define $U := -U_\psi U_g$ and calculate its action in the hyperplane $\mathsf{H}_\psi := \mathrm{span}\{|\psi_g\rangle, |\psi_b\rangle\}$:

$$U|\psi_g\rangle = \cos(2\theta)|\psi_g\rangle - \sin(2\theta)|\psi_b\rangle, \qquad U|\psi_b\rangle = \cos(2\theta)|\psi_b\rangle + \sin(2\theta)|\psi_g\rangle.$$

U acts as a rotation of 2θ in the hyperplane H_ψ, so after k repeated applications of U on $|\psi\rangle$, the initial state evolves in:

$$U^n|\psi\rangle = \sin[(2k+1)\theta]|\psi_g\rangle + \cos[(2k+1)\theta]|\psi_b\rangle. \tag{4.3.4}$$

The probability to obtain a target item by means of a measurement on the resulting state is $\mathbb{P} = \sin^2[(2k+1)\theta]$. Since $\mathbb{P} = 1$ for $k = \frac{\pi}{4\theta} - \frac{1}{2}$, the number of applications of U that maximizes $\mathbb{P}$ is $\lfloor\frac{\pi}{4\theta}\rfloor$. Therefore, U is a generalized Grover's iteration.

If there are M target items over N items in the database and the initial state is the superposition of all the items $|\psi\rangle = \frac{1}{\sqrt{N}}\sum_{x\in X}|x\rangle$ then the norm of the projection of $|\psi\rangle$ onto H_g is:

$$\sin\theta = \left\|\sum_{x\in T}|x\rangle\langle x|\psi\rangle\right\| = \frac{1}{\sqrt{N}}\left\|\sum_{x\in T}|x\rangle\right\| = \sqrt{\frac{M}{N}}.$$

Therefore, assuming $M \ll N$, $\theta \simeq \sin\theta$ so the number of required iterations is $k = \lfloor\frac{\pi}{4}\sqrt{\frac{N}{M}}\rfloor$.

Without the knowledge of the number M of solutions marked by the oracle, one cannot use amplitude amplification as it is. When M is not know a priori, one can use the BBHT algorithm [BBHT98] to find a solution in a set of N items given an oracle that marks the solutions. Assuming $1 \leq M \leq 3N/4$, the BBHT algorithm is given by Algorithm 1 that finds a solution in time $O(\sqrt{N/M})$.

Input: Unsorted database X of N items, oracle function $f : X \to \{0,1\}$ marking the solutions
Result: Solution
1   initialize $m = 1$ and set $\lambda \in [1, 4/3]$;
2   select r uniformly at random in $\{a \in \mathbb{N} : a \leq m\}$;
3   apply r iterations of amplitude amplification with initial state $|\psi\rangle = \frac{1}{\sqrt{N}}\sum_{x\in X}|x\rangle$);
4   measure the register (let $\hat{x}$ be the outcome);
5   **if** $f(\hat{x}) = 1$ **then**
6 | **return** $\hat{x}$
7   **else**
8 | set m to $\min(\lambda m, \sqrt{N})$ and go back to line 2;
9   **end**

Algorithm 1: *BBHT algorithm*

Algorithm 1 can be adapted to define a remarkable algorithm for finding the minimum of a function, the so-called *Dürr-Høyer algorithm* [DH96]. Let $F : \{0, ..., N -$

$1\} \to \mathbb{R}$ be a real function over a set of N items and define the oracle function:

$$f_y(x) := \begin{cases} 1 & F(x) < y \\ 0 & F(x) \geq y. \end{cases} \qquad (4.3.5)$$

with a threshold $y \in \mathrm{Im}(f)$ for the amplitude amplification. The Dürr-Høyer algorithm is given in Algorithm 2.

Input: Function $F : \{0, ..., N-1\} \to \mathbb{R}$ to minimize
Result: $\mathrm{argmin} F$
1 select x uniformly at random and set a threshold $y = f(x)$;
2 **Repeat** $O(\sqrt{N})$ **times**
3   apply BBHT algorithm with oracle function f_y;
4   measure the register (let $\hat{x}$ be the outcome);
5   **If** $F(\hat{x}) < F(x)$ **then** $x \leftarrow \hat{x}$;
6 **return** x

Algorithm 2: *Dürr-Høyer algorithm.*

4.4 Quantum phase estimation

Given a unitary operator U on the Hilbert space H and an eigenvector of U that is $|\psi\rangle \in \mathsf{H}$ such that $U|\psi\rangle = e^{2\pi i\theta}|\psi\rangle$ for some $\theta \in [0,1)$, the quantum phase estimation (QPE) algorithm returns an estimation of the phase θ [CN00]. A run of QPE is described by the following circuit:

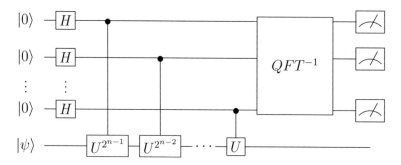

where there are n ancillary qubits initialized in $|0\rangle$ and the main register initialized in $|\psi\rangle$. The total initial state is $|\Psi\rangle_{in} = |0\rangle^{\otimes n} \otimes |\psi\rangle$. The action of the n Hadamard gates on the ancillas produces the state:

$$(H^{\otimes n} \otimes \mathbb{I})|\Psi\rangle_{in} = \frac{1}{2^{n/2}}(|0\rangle + |1\rangle)^{\otimes n} \otimes |\psi\rangle.$$

Since $U^{2^j}|\psi\rangle = e^{2\pi i 2^j \theta}|\psi\rangle$, we have that, after the action of the controlled gates, the state of the register is:

$$|\Psi\rangle = \frac{1}{\sqrt{2}}\left(|0\rangle + e^{2\pi i 2^{n-1}\theta}|1\rangle\right)\otimes\cdots\otimes\frac{1}{\sqrt{2}}\left(|0\rangle + e^{2\pi i\theta}|1\rangle\right)\otimes|\psi\rangle = \frac{1}{2^{n/2}}\sum_{k=0}^{2^n-1}e^{2\pi ik\theta}|k\rangle\otimes|\psi\rangle.$$

The action of the inverse quantum Fourier transform is:

$$(QFT^{-1}\otimes\mathbb{I})|\Psi\rangle = \frac{1}{2^n}\sum_{x=0}^{2^n-1}\sum_{k=0}^{2^n-1}e^{-\frac{2\pi ik}{2^n}(x-2^n\theta)}|x\rangle\otimes|\psi\rangle \equiv |\Psi\rangle_f.$$

Let y be the nearest integer to $2^n\theta$, so $2^n\theta = y + 2^n\delta$ with $|2^n\delta| \leq 1/2$:

$$|\Psi\rangle_f = \frac{1}{2^n}\sum_{x=0}^{2^n-1}\sum_{k=0}^{2^n-1}e^{-\frac{2\pi ik}{2^n}(x-y)}e^{2\pi ik\delta}|x\rangle\otimes|\psi\rangle, \tag{4.4.1}$$

the probability to obtain y measuring the n ancillary qubits is:

$$\mathbb{P}_{|\Psi\rangle_f}(y) = \left|\langle y|\frac{1}{2^n}\sum_{x=0}^{2^n-1}\sum_{k=0}^{2^n-1}e^{-\frac{2\pi ik}{2^n}(x-y)}e^{2\pi ik\delta}|x\rangle\right|^2$$

$$= \left|\frac{1}{2^n}\sum_{x=0}^{2^n-1}\sum_{k=0}^{2^n-1}e^{-\frac{2\pi ik}{2^n}(x-y)}e^{2\pi ik\delta}\delta_{xy}\right|^2$$

$$= \left|\frac{1}{2^n}\sum_{k=0}^{2^n-1}e^{2\pi ik\delta}\right|^2$$

$$= \frac{1}{2^{2n}}\left|\frac{1-e^{2\pi i 2^n\delta}}{1-e^{2\pi i\delta}}\right|^2 \geq \frac{4}{\pi^2} \simeq 0.41 \quad\text{for}\quad \delta\neq 0. \tag{4.4.2}$$

If $2^n\theta$ is an integer number, that is $\delta = 0$, then $\mathbb{P}_{|\Psi\rangle_f}(y = 2^n\theta) = 1$. As explicitly shown in [CN00], $\mathbb{P}_{|\Psi\rangle_f}(y) \geq 1 - \epsilon$ using $O(\log(1/\epsilon))$ ancillary qubits, corresponding to $O(\epsilon^{-1})$ controlled gates U.

Let us consider the case that the main register is not initialized in $|\psi\rangle$ but in another state $|\varphi\rangle$ that is not an eigenstate of U. Let $\{|\psi_i\rangle\}$ be the orthonormal basis of H made by the eigenstates of U with eigenvalues $e^{2\pi i\theta_i}$, so $|\varphi\rangle = \sum_i a_i|\psi_i\rangle$. The output $|\Psi\rangle_f$ of QPE is a state close to:

$$|\phi\rangle = \sum_i a_i|y_i\rangle\otimes|\psi_i\rangle, \tag{4.4.3}$$

in the sense that a measurement on $|\Psi\rangle_f$ provides an estimation of the output of a measurement on $|\phi\rangle$ as shown by (4.4.1). Therefore, using $O(\log(1/\epsilon))$ ancillary qubits, the probability to find the nearest integer y to $2^n\theta$ satisfies $\mathbb{P}(y) \geq |a_\psi|^2(1-\epsilon)$, where a_ψ is the coefficient of $|\psi\rangle$ within the superposition $|\phi\rangle$.

In the following, we will apply QPE algorithm as a subroutine in several quantum machine learning algorithms like quantum principal component analysis, quantum support vector machine, and quantum perceptron.

Chapter 5

QML toolkit

Once described some quantum algorithms that are important to devise quantum machine learning (QML) schemes, in this chapter we list some notions and technical tools that are widely used in QML. The quantum random access memory (QRAM) is a model for the efficient retrieval of quantum states encoding data, a procedure of Hamiltonian simulation is crucial for some QML algorithms that we describe in the following, the SWAP test and the quantum calculation of distances are important tools adopted in several quantum algorithms for classification and clustering.

5.1   QRAM

Data encoded into quantum states must be efficiently retrieved in order to be processed by a quantum machine. This requirement is particularly crucial in QML where one assumes to have huge amounts of data stored in quantum memories. The quantum version of the random access memory (RAM) was proposed as a model for data retrieval in quantum computing [Gi08]. A classical RAM is formed by an array of memory cells, an input register to address the cells and an output register to return the information stored in the memory. A quantum random access memory (QRAM) is basically the same concept where the address and output registers are realized by qubits, while the array of memory cells may be both classical or quantum depending on hardware architectures. The input register of a QRAM can be in a superposition of queryings for N different addresses $\sum_{i=1}^{N} \alpha_i |i\rangle_{in} \otimes |0\rangle_{out}$, in this case the output register returns a superposition of stored data $\sum_{i=1}^{N} \alpha_i |i\rangle_{in} \otimes |\psi_i\rangle_{out}$ where $|\psi_i\rangle$ represents the content of the i-th memory location.

Let us consider the architecture of a classical RAM given by a decision tree like in Figure 5.1. Each bit in the address register encodes the left/right direction at any level toward the target memory cell. From the viewpoint of physical implementation,

© The Author(s), under exclusive license to Springer Nature Singapore Pte Ltd. 2023
D. Pastorello, *Concise Guide to Quantum Machine Learning*,
Machine Learning: Foundations, Methodologies, and Applications,
https://doi.org/10.1007/978-981-19-6897-6_5

in any node there is an activated transistor that deviates a signal according to the input address. Building directly a QRAM with this architecture is impractical because it is difficult maintain the necessary quantum coherence in the last graph levels.

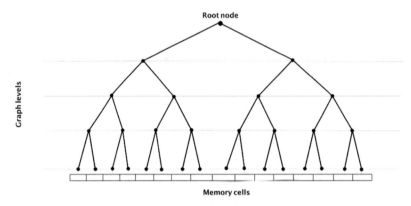

Figure 5.1. A bifurcation tree composed by the paths to 16 memory cells of a RAM. The k-th bit in the address register controls the activation of the 2^k bifurcation nodes at the k-th graph level.

In [Gi08] there is a proposal to implement an effective QRAM by means of the *bucket brigade* architecture reducing the number of nodes activations per memory call. The nodes of the bifurcation tree are *qutrits* (3-level systems) with ground state $|wait\rangle$ and two excited states $|0\rangle$ and $|1\rangle$. Consider a n-qubit input register, the 2^n nodes of the tree are initially prepared in $|wait\rangle$. Let $|a_0 a_1 \cdots a_{n-1}\rangle$ be an address specified in the input register, the 0-th qubit is input at the root of the tree interacting with the qutrit at level 0 inducing the transition $|wait\rangle \to |a_0\rangle$. If the root qutrit is in $|0\rangle$ then the following address qubit is routed to the left, if the quitrit is in $|1\rangle$ then the address qubit is routed to the right in the tree. The address qubit in $|a_1\rangle$ interacts with the qutrit in the root node and it is routed to a node at the first level inducing the transition $|wait\rangle \to |a_1\rangle$ of the corresponding qutrit. Every address qubit is sent through the tree, the k-th address qubit changes the state of a single node at the k-th level. At the end of the process n qutrits are in a excited state (activated nodes) giving rise to a routing path to the target memory location. In the case of a superposition of addresses the result is a superposition of routes with $O(n)$ activated nodes. When the routing path has been realized, a *bus qubit* is sent to interact with the target memory cells, then it is sent back to write the result on the output register. To retrieve a state with the QRAM, $O(n)$ operations, corresponding to nodes activations, are needed.

A proposed physical implementation of QRAM uses trapped ions as routing nodes and polarized photons as addressing qubits [Gi08]. A more recent proposal presents a *circuit-based flip-flop QRAM* to construct a quantum database such that for registering or updating classical data consisting of N entries, each represented by n bits, the method requires $O(n)$ qubits and $O(Nn)$ steps [PPR19].

5.2 Hamiltonian simulation

The problem of *Hamiltonian simulation* is general and well-known in quantum mechanics and quantum information theory. Its statement can be summarized as follows: given a Hamiltonian $\mathcal{H}$, find a quantum algorithm to implement a unitary operator U such that $\|U - e^{-i\mathcal{H}\Delta t}\|_{op} < \epsilon$ for a given time interval Δt and a maximum error $\epsilon > 0$.

In view of our purposes, we focus on a specific Hamiltonian simulation problem. Let us consider a quantum state described by the density matrix ρ on the Hilbert space H. Since ρ is a self-adjoint operator by definition, the operator $U = e^{-i\rho t}$ is unitary and describes the time evolution of a quantum system under the action of ρ as the Hamiltonian. More precisely, consider two copies S_1 and S_2 of a quantum system that are described in $\mathsf{H} \otimes \mathsf{H}$, S_1 is prepared in the state ρ and S_2 is prepared in the state σ. We need to operate over the systems S_1 and S_2 so that the output state of S_2 is $\sigma' = e^{-i\rho\Delta t}\sigma$ for a fixed Δt. In this case, we simulate the Hamiltonian ρ whose peculiarity is that its physical meaning is not an *energy* but a state.

In order to define a simulation procedure of the Hamiltonian ρ, let us consider the swap operator S on $\mathsf{H} \otimes \mathsf{H}$ whose action on the product vectors is $S|\psi\varphi\rangle = |\varphi\psi\rangle$. For instance, in the case of a qubit pair, the swap operator on $\mathbb{C}^2 \otimes \mathbb{C}^2$ (that is the SWAP gate introduced in (3.2.5) is given by:

$$S = \frac{1}{2}(\mathbb{I}_{\mathbb{C}^4} + \sigma_x \otimes \sigma_x + \sigma_y \otimes \sigma_y + \sigma_z \otimes \sigma_z) = \begin{pmatrix} 1 & 0 & 0 & 0 \\ 0 & 0 & 1 & 0 \\ 0 & 1 & 0 & 0 \\ 0 & 0 & 0 & 1 \end{pmatrix}. \qquad (5.2.1)$$

Since S corresponds to a sparse matrix, we have that the swap operator is a Hamiltonian that can be easily implemented or simulated.

***Proposition* 5.2.1** *Let H be a Hilbert space. Let S be the swap operator on $\mathsf{H} \otimes \mathsf{H}$ and ρ, σ be density matrices on H. The following identity is true:*

$$\mathrm{tr}_2[e^{-iS\Delta t}(\sigma \otimes \rho)e^{iS\Delta t}] = \sigma - i\Delta t[\rho, \sigma] + o(\Delta t^2), \qquad (5.2.2)$$

where tr_2 is the partial trace over the second tensor factor.

Proof. Considering the Taylor's series of the exponential function and taking equation (2.2.10) into account, we have:

$$\text{tr}_2[e^{-iS\Delta t}(\sigma \otimes \rho)e^{iS\Delta t}] = \text{tr}_2\left[\sum_{k,k'=0}^{\infty} \frac{(-i)^k \Delta t^k i^{k'} \Delta t^{k'}}{k! k'!} S^k(\sigma \otimes \rho)S^{k'}\right].$$

Assume $\Delta t \ll 1$ and let $\rho = \sum_{i,i'} \rho_{ii'}|i\rangle\langle i'|$ and $\sigma = \sum_{j,j'} \sigma_{jj'}|j\rangle\langle j'|$ be the considered density matrices expressed in diagonal form in terms of $\{|i\rangle\}_i$ and $\{|j\rangle\}_j$ that are orthonormal bases of H_d. Therefore:

$$\text{tr}_2[e^{-iS\Delta t}(\sigma \otimes \rho)e^{iS\Delta t}] = \text{tr}_2\left[\left(\sum_{ii'}\sum_{jj'}\sigma_{jj'}\rho_{ii'}|j\rangle\langle j'| \otimes |i\rangle\langle i'|\right) + \right.$$

$$+i\Delta t\left(\sum_{ii'}\sum_{jj'}\sigma_{jj'}\rho_{ii'}|j\rangle\langle j'| \otimes |i\rangle\langle i'|\right)S-$$

$$\left.-i\Delta tS\left(\sum_{ii'}\sum_{jj'}\sigma_{jj'}\rho_{ii'}|j\rangle\langle j'| \otimes |i\rangle\langle i'|\right) + o(\Delta t^2)\right].$$

Applying the SWAP operator and tracing out the second system:

$$\text{tr}_2[e^{-iS\Delta t}(\sigma \otimes \rho)e^{iS\Delta t}] = \sum_{jj'}\sigma_{jj'}|j\rangle\langle j'| + i\Delta t\sum_{i'j}\sum_{q}\sigma_{jq}\rho_{qi'}|j\rangle\langle i'|-$$

$$-i\Delta t\sum_{ij'}\sum_{q}\sigma_{qj'}\rho_{iq}|i\rangle\langle j'| + o(\Delta t^2).$$

Using $\langle j'|i\rangle = \delta_{j'i}$ and $\langle i'|j\rangle = \delta_{i'j}$, we get (5.2.2):

$$\text{tr}_2[e^{-iS\Delta t}(\sigma \otimes \rho)e^{iS\Delta t}] = \sigma + i\Delta t\sum_{i'j}\sum_{ij'}\rho_{ii'}\sigma_{jj'}|j\rangle\langle j'|i\rangle\langle i'|-$$

$$-i\Delta t\sum_{ij'}\sum_{i'j}\rho_{ii'}\sigma_{jj'}|i\rangle\langle i'|j\rangle\langle j'| + o(\Delta t^2) =$$

$$= \sigma - i\Delta t[\rho, \sigma] + o(\Delta t^2).$$

$\square$

We combine equation (5.2.2) with the following:

$$e^{-i\rho\Delta t}\sigma e^{i\rho\Delta t} = (1 - i\rho\Delta t + o(\Delta t^2))\sigma(1 + i\rho\Delta t + o(\Delta t^2))$$

$$= \sigma - i\Delta t\rho\sigma + i\Delta t\sigma\rho + o(\Delta t^2)$$

$$= \sigma - i\Delta t[\rho,\sigma] + o(\Delta t^2).$$

Therefore, we have that:

$$e^{-i\rho\Delta t}\sigma e^{i\rho\Delta t} = \text{tr}_2[e^{-iS\Delta t}(\sigma \otimes \rho)e^{iS\Delta t}] + o(\Delta t^2). \qquad (5.2.3)$$

Equation (5.2.3) gives the prescription for simulating the dynamics under the action of the Hamiltonian ρ. Given two copies of the considered system prepared in $\sigma \otimes \rho$, the reduced dynamics of the first system for a short time interval Δt under the action of the Hamiltonian S realizes an effective simulation of the dynamics under the action of the Hamiltonian given by ρ. Repeated applications of $e^{-iS\Delta t}$, using multiple copies of ρ, allows to simulate $e^{-i\rho t}$ with $t = n\Delta t$. In particular, by means of the described Hamiltonian simulation, once retrieved a state ρ from a QRAM we are able to implement the unitary operator $e^{-i\rho t}$, in other words we can exponentiate density matrices. For example, the exponentiation procedure of a density matrix allows to obtain its eigendecomposition applying the phase estimation algorithm. This is the key idea of the quantum principal component analysis where a covariance matrix is encoded into a density matrix.

5.3 SWAP test

The SWAP test was introduced in [Bu01] to estimate the real number $|\langle\psi|\varphi\rangle|^2$, given two unknown pure states $|\psi\rangle$ and $|\varphi\rangle$. The procedure requires a controlled swap operation, between two multiple qubit registers, generalizing the Fredkin gate introduced in (3.2.6). It is straightforward to show that the controlled SWAP of two n-qubit registers can be implemented with n Fredkin gates controlled by the same qubit.

Consider two n-qubit registers initialized in $|\psi\rangle$ and $|\varphi\rangle$ and an ancillary qubit prepared in $|0\rangle$. Assume to act with the following circuit:

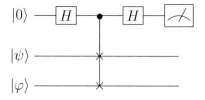

Denoting the action of the controlled SWAP by F, the circuit returns the state:

$$|\Psi\rangle = (H \otimes \mathbb{I} \otimes \mathbb{I})\mathsf{F}(H \otimes \mathbb{I} \otimes \mathbb{I})|0\,\psi\,\varphi\rangle = (H \otimes \mathbb{I} \otimes \mathbb{I})\mathsf{F}|+\psi\,\varphi\rangle =$$

$$= (H \otimes \mathbb{I} \otimes \mathbb{I})\frac{1}{\sqrt{2}}(|0\,\psi\,\varphi\rangle + |1\,\varphi\,\psi\rangle) = \frac{1}{2}|0\rangle(|\psi\varphi\rangle + |\varphi\psi\rangle) + \frac{1}{2}|1\rangle(|\psi\varphi\rangle - |\varphi\psi\rangle).$$

The probability to measure 0 on the first qubit is:

$$\mathbb{P}(0) = \langle\Psi|(|0\rangle\langle0| \otimes \mathbb{I} \otimes \mathbb{I})\Psi\rangle = \frac{1}{4}((\langle\psi\,\varphi| + \langle\varphi\,\psi|)(|\psi\,\varphi\rangle + |\varphi\,\psi\rangle)) =$$

$$= \frac{1}{2} + \frac{1}{4}(\langle\psi|\varphi\rangle\langle\varphi|\psi\rangle + \langle\varphi|\psi\rangle\langle\psi|\varphi\rangle) = \frac{1}{2} + \frac{1}{2}|\langle\psi|\varphi\rangle|^2.$$

Therefore, if we have two unknown states in input, repeated measurements on the ancillary qubit allow to estimate the squared modulus of their inner product. Despite the simple idea behind it, the SWAP test is a useful tool for machine learning applications, in particular for implementing distance-based clustering and classification.

5.4 Qdist routine

The SWAP test can be applied directly for calculating dot products and distances among real vectors within the amplitude encoding. In this section we outline a procedure, that we call Qdist routine, to efficiently estimate the Euclidean distance in logarithmic time assuming that the real components of the considered vectors and their norms are stored as floating point numbers in a QRAM.

Let $\mathbf{x} \in \mathbb{R}^d$ be encoded into the amplitudes of a quantum state in a $\log d$-qubit register:

$$|\mathbf{x}\rangle = \frac{1}{\|\mathbf{x}\|}\sum_{j=1}^{d}\mathbf{x}_j|j\rangle. \tag{5.4.1}$$

Observe that the dot product between two real vectors is related to the inner product between the corresponding quantum states by the cosine similarity:

$$\langle\mathbf{x}|\mathbf{y}\rangle = \cos(\mathbf{x}, \mathbf{y}) = \frac{\mathbf{x} \cdot \mathbf{y}}{\|\mathbf{x}\|\|\mathbf{y}\|}. \tag{5.4.2}$$

Consider two real vectors $\mathbf{x}, \mathbf{y} \in \mathbb{R}^d$ and their amplitude encodings $|\mathbf{x}\rangle$ and $|\mathbf{y}\rangle$ in a n-qubit register with $n = \log d$, the SWAP test can be applied to compute their dot product. The procedure starts with the preparation of the states (considering two ancillary qubits, called *ancilla1* and *ancilla2*):

$$|\psi\rangle = \frac{1}{\sqrt{2}}(|0\rangle \otimes |\mathbf{x}\rangle + |1\rangle \otimes |\mathbf{y}\rangle) \in \mathbb{C}^2 \otimes (\mathbb{C}^2)^{\otimes n} \tag{5.4.3}$$

$$|\varphi\rangle = \frac{1}{\sqrt{2}}(|0\rangle - |1\rangle) \in \mathbb{C}^2. \tag{5.4.4}$$

In (5.4.3) we have an ancillary qubit that is entangled with the n-qubit register where the balanced superposition of $|\mathbf{x}\rangle$ and $|\mathbf{y}\rangle$ is stored. Assuming that the real components of $\mathbf{x}$ and $\mathbf{y}$ are stored in an array of memory cells as floating point numbers and the norms are given separately, the retrieval of the state (5.4.3) can be done in time $O(\log d)$ from the QRAM. The non-normalized vector[1] obtained taking the inner product between the state (5.4.4) and the tensor factors corresponding to the ancillary qubit in the state (5.4.3) is:

$$\Phi = \frac{1}{\sqrt{2}}(\langle\varphi|0\rangle|\mathbf{x}\rangle + \langle\varphi|1\rangle|\mathbf{y}\rangle) = \frac{1}{\sqrt{2}}(|\mathbf{x}\rangle - |\mathbf{y}\rangle). \tag{5.4.5}$$

The SWAP test performed on the two ancillas allows to estimate $\| \Phi \|^2$. The fact can be easily explained as follows: Let $|\psi_1\rangle \otimes |\psi_2\rangle \in \mathsf{H}_1 \otimes \mathsf{H}_2$ and $|\varphi\rangle \in \mathsf{H}_1$, take the vector $\phi := \langle\varphi|\psi_1\rangle|\psi_2\rangle \in \mathsf{H}_2$ and observe that $\| \phi \|^2 = \langle\phi|\phi\rangle = |\langle\varphi|\psi_1\rangle|^2$. This argument extends by linearity to the superposition of (5.4.5). The importance of estimating $\| \Phi \|^2$ is motivated by the following:

$$\| \Phi \|^2 = 1 - \frac{\mathbf{x} \cdot \mathbf{y}}{\| \mathbf{x} \| \| \mathbf{y} \|}. \tag{5.4.6}$$

Since the norms of the vectors must be stored separately, as required by the amplitude encoding, the estimation of $\| \Phi \|^2$ allows to estimate

$$\mathbf{x} \cdot \mathbf{y} = (1 - \| \Phi \|^2) \| \mathbf{x} \| \| \mathbf{y} \|,$$

and consequently the Euclidean distance $\| \mathbf{x} - \mathbf{y} \|^2 = \| \mathbf{x} \|^2 + \| \mathbf{y} \|^2 - 2\mathbf{x} \cdot \mathbf{y}$.

The number of repeated measurements, required to run the SWAP test, controls the accuracy in the estimation of the distance but it does not depend on the dimension of the space where the distance is calculated. Let us estimate the necessary number of algorithm runs: Let $\mathbb{P}$ be the probability to obtain the outcome 1 (success), R be the number of runs and $l_1, ..., l_R$ be the collected outcomes. Repeated

---

[1]The vector Φ does not present a physical meaning by itself, it is just an algebraic object not a quantum state, for this reason we omit the ket symbol.

runs correspond to a sampling from a Bernoulli distribution and we can calculate the *binomial proposition confidence interval* for the probability $\mathbb{P}$ of success [Br01, Sc17]:

$$I = \hat{\mathbb{P}} \pm z\sqrt{\frac{\hat{\mathbb{P}}(1-\hat{\mathbb{P}})}{R}}, \tag{5.4.7}$$

where $\hat{\mathbb{P}}$ is the proportion of successes and z is a confidence level expressed as a value of the quantile function associated with the normal distribution[2]. The maximum error $z\sqrt{\frac{\hat{\mathbb{P}}(1-\hat{\mathbb{P}})}{R}}$ is reached for $\hat{\mathbb{P}} = 0.5$ so the estimation error ϵ is at most $\frac{z}{2\sqrt{R}}$ with a confidence level of z. Thus, the number R of repetitions required for an estimation within an error $\epsilon > 0$ is $O(\epsilon^{-2})$. The complexity of the calculation of the Euclidean distance between two vectors in $\mathbb{R}^d$ by means of the **Qdist** routine, up to an error $\epsilon > 0$, is $O(\epsilon^{-2} \log d)$.

Qdist is called as a subroutine in several quantum clustering algorithms, described in chapter 6, for example quantum K-means is based on a generalization of **Qdist** combined with a quantum calculation of the cluster centroids. Moreover, modified versions of **Qdist** are used for quantum classification algorithms as discussed in chapter 7.

---

[2]For example, the value corresponding to 99% is $z = 2.58$

Chapter 6

Quantum clustering

In unsupervised learning, algorithms are not guided by labels of the training set but they must identify structures into data instances, for example one would like to find preferred directions in the feature space to identify geometric insights or performing a dimensionality reduction to dimension 2 or 3 for providing a graphical representation of the dataset. Without labels, there are not known classes of data a priori but a learning algorithm must be able to divide the data instances into clusters formed by similar data. In this section we present the quantum principal component analysis for dimensionality reduction of the feature space and some clustering techniques obtained embedding quantum subroutines into K-means, K-medians and divise clustering algorithms. Moreover, we discuss the implementation of a 2-means clustering with a quantum annealer.

6.1 Quantum principal component analysis

Given a dataset represented in a features space, *principal component analysis* (PCA) is an algorithm to find the direction of maximum variance in the dataset, the *first principal component*, and the orthogonal directions, the *further principal components*, in decreasing order of variance. The result is a new basis of the feature space that diagonalizes the covariance matrix, then the corresponding features are uncorrelated. PCA allows to select directions of low variance that can be suppressed representing data in a space with lower dimension. PCA is a method of feature extraction since it provides new uncorrelated features by means of a change of basis, it is also a method of feature selection because it allows to reduce the number of effective features.

Let $\mathcal{F}$ be a real d-dimensional feature space, so each feature of the considered phenomenon is identified with a vector of the standard basis $\{\mathbf{e}_i\}_{i=1,\dots,d}$ of $\mathbb{R}^d$. Given

© The Author(s), under exclusive license to Springer Nature Singapore Pte Ltd. 2023   57
D. Pastorello, *Concise Guide to Quantum Machine Learning*,
Machine Learning: Foundations, Methodologies, and Applications,
https://doi.org/10.1007/978-981-19-6897-6_6

a dataset of N instances, the i-th instance is represented by a vector in the feature space $\mathbf{x}_i = \sum_{j=1}^{d} x_{ij}\mathbf{e}_i$ for any $i = 1, ..., N$. The $N \times d$ matrix $X := (x_{ij})$ is called *data matrix* and provides the description of the considered dataset into the feature space of reference: the rows of X corresponds to the data instances and the columns corresponds to the features. Let X be a $N \times d$ data matrix where the values of the features are rescaled so that the mean over the columns is zero that is $\mathbb{E}_i(x_{ij}) = 0$ for any $j = 1, ..., d$. Let us define the *covariance matrix* as $M := X^T X$ that is a symmetric $d \times d$ matrix. Up to the factor N^{-1}, the element $m_{kk} = \sum_{i=1}^{N} x_{ik}^2$ is the variance of the k-th feature and the off-diagonal element $m_{kl} = \sum_{i=1}^{N} x_{ik}x_{il}$ is the covariance between the k-th feature and the l-th feature.

Let W be the matrix of eigenvectors of M, so:

$$M = W\Lambda W^T, \tag{6.1.1}$$

where Λ is the matrix of eigenvalues. Assume that the eigenvalues on the diagonal of Λ are arranged in decreasing order so the eigenvectors in W are accordingly arranged. We define the $N \times d$ matrix P by projecting X onto the eigenvectors basis:

$$P := XW. \tag{6.1.2}$$

P is the transformed data matrix where the greatest variance of the data lies on the first coordinates of data vectors, so the last coordinates may be discarded because of their small variance that is low representativity in the dataset. In other words, the direction of maximum variance in the dataset is individuated by the eigenvector of M with higher eigenvalue. Given a dataset of N items described by d features, classical PCA requires $O(dN^2)$ operations to generate the covariance matrix and $O(d^3)$ operations for its eigendecomposition. Since the full eigendecomposition may be not necessary, there are effective techniques to calculate the greatest k eigenvalues and relative eigenvectors in time $O(kdN)$.

The quantum version of PCA is based on the preparation of a density matrix ρ representing the (normalized) covariance matrix of a dataset. Then the quantum phase estimation algorithm is applied to the unitary operator $e^{-i\rho t}$, that must be efficiently simulated, to find eigenvalues and eigenvectors of ρ [Ll14]. The first goal is encoding a given covariance matrix M into a quantum state, more precisely we need to represent M as a density matrix of a quantum system. Let us define the normalized version of M with unit trace $M_1 := \frac{1}{C}M$ where $C = \sum_{k=1}^{d} \sum_{i=1}^{N} x_{ik}^2 = \sum_i \|\mathbf{x}_i\|^2$. Let $\mathbf{x}_i$ be a data point in the feature space and $|\mathbf{x}_i\rangle = \|\mathbf{x}_i\|^{-1} \sum_{k=1}^{d} x_{ik}|k\rangle \in \mathsf{H}(\simeq \mathbb{C}^d)$ be the corresponding quantum state within the amplitude encoding. The data matrix can be written in these terms:

$$X = \sum_{i=1}^{N} \|\mathbf{x}_i\| |i\rangle\langle \mathbf{x}_i|, \tag{6.1.3}$$

where $\{|i\rangle\}_{i=1,\dots,N}$ is an orthonormal basis of $\mathbb{C}^N$ that can be identified as a Hilbert space of a $\log N$-qubit register that we use to index the data instances. Therefore the covariance matrix is:

$$M = X^T X = \sum_{i,j} \|\mathbf{x}_i\| \|\mathbf{x}_j\| |\mathbf{x}_i\rangle\langle i|j\rangle\langle \mathbf{x}_j| = \sum_i \|\mathbf{x}_i\|^2 |\mathbf{x}_i\rangle\langle \mathbf{x}_i|,$$

since $\langle i|j\rangle = \delta_{ij}$. Let us remark that M is not diagonalized in this form because $\{|\hat{\mathbf{x}}_i\rangle\}_i$ is not an orthonormal basis. If we define the following pure state of the composite system described in $\mathbb{C}^d \otimes \mathbb{C}^N$:

$$|\Psi\rangle := \frac{1}{\sqrt{C}} \sum_{i=1}^{N} \|\mathbf{x}_i\| |\mathbf{x}_i\rangle |i\rangle, \tag{6.1.4}$$

we have that M_1 is obtained tracing out the second subsystem from $|\Psi\rangle$:

$$M_1 = \rho = \mathrm{tr}_{\mathbb{C}^N}(|\Psi\rangle\langle\Psi|). \tag{6.1.5}$$

Given a dataset of N instances described in a d-dimensional feature space, the physical implementation of the associated covariance matrix as a state of a quantum system is based on the amplitude encoding of any data point into a pure state of a d-level quantum system. This system is entangled with another quantum system with Hilbert space of dimension N by the preparation of the state (6.1.4). The reduced state ρ of the first subsystem is the mixed state representing the covariance matrix. To retrieve the state ρ with a QRAM, $O(\log(Nd))$ operations are needed (see section 5.1).

To find eigenvectors and eigenvalues of ρ we can apply the quantum phase estimation algorithm on $e^{-i\rho t}$, for varying time, that is the unitary operator describing the time evolution of a d-level quantum system under the action of ρ as the Hamiltonian that can be simulated as described in Section (5.2). The simulation of the time evolution operator $e^{-i\rho t}$ to accuracy ϵ requires $n = O(t^2\epsilon^{-1} \| \rho - \sigma \|) < O(t^2\epsilon^{-1})$ steps, where $t = n\Delta t$ [Ll14]. Finally, quantum phase estimation (QPE) is run using conditional applications of $e^{-i\rho t}$ for varying times, taking ρ itself as initial state. QPE returns the state:

$$\sum_i \lambda_i \left|\hat{\lambda}_i\right\rangle\left\langle\hat{\lambda}_i\right| \otimes |\psi_i\rangle\langle\psi_i|, \tag{6.1.6}$$

where λ_i are the eigenvalues of ρ, $\hat{\lambda}_i$ are the estimations of the eigenvalues provided by QPE and $|\psi_i\rangle$ are the eigenvectors of ρ. QPE returns the eigenvectors and eigenvalues to accuracy ϵ in time $O(\epsilon^{-1})$. The complexity of QPCA grows logarithmically in Nd, this is an exponential speed-up with respect to any known classical algorithm. However, the QPCA algorithm described in this section requires the availability of a QRAM for data retrieval and the implementation of QPE, therefore there is still a lack of an experimental demonstration.

6.2   Quantum K-means

The K-means algorithm is a method for partitioning a set of data points into K clusters so that the variance is minimized within each cluster. Let V be a dataset of N instances described in the d-dimensional feature space $\mathcal{F}$. Let $P = \{P_1, ..., P_K\}$ be a partition of V into K clusters that is a collection of subsets of V satisfying:

$$P_i \cap P_j = \emptyset \quad \text{for} \quad i \neq j \quad \text{and} \quad \bigcup_{j=1}^{K} P_j = V.$$

The *centroid* of P_j is defined as the mean of its elements:

$$\mathbf{c}_j := \frac{1}{N_j} \sum_{\mathbf{x} \in P_j} \mathbf{x} \quad \text{where} \quad N_j = |P_j|. \tag{6.2.1}$$

The K-means clustering is outlined in Algorithm 3: It starts from a random partition of the dataset, then the centroids of the clusters are calculated and each data point is attached to the nearest centroid obtaining a new partition. The process is iterated until convergence, that is, no new partitions are generated. The algorithm minimizes the following objective function f defined on all the partitions of V into K clusters:

$$f(P) = \sum_{j=1}^{K} \sum_{\mathbf{x} \in P_j} \mathsf{d}(\mathbf{x}, \mathbf{c}_j), \tag{6.2.2}$$

where d is the reference distance defined on $\mathcal{F}$. Each iteration requires $O(NKd)$ operations and the number of required iterations to convergence is often less than N as an empirical evidence. In general the algorithm returns just a local minimum of f.

A quantum version of the K-means algorithm allows an efficient calculation of the distances between each data instance and each centroid by means of the Qdist routine. Let us assume $\mathcal{F} = \mathbb{R}^d$ equipped with the Euclidean distance. A

Input: Data set V, number of clusters K
Result: Partition of V into K clusters
1 select an initial partition P of V;
2 **repeat**
3   calculate the centroid $\mathbf{c}_j$ of any $P_j \in P$;
4   construct a new partition P' whose clusters are defined by:
      $P'_j := \{\mathbf{x} \in V : \mathbf{c}_j \text{ is the nearest centroid}\} \quad \forall j = 1, ..., K$;
5 **until** *convergence*;
6 **return** P'

Algorithm 3: *K-means clustering algorithm.*

feature vector can be represented into a n-qubit register by means of the amplitude encoding, without loss of generality let us assume that $d = 2^n$. Let $\{\mathbf{x}_1, ..., \mathbf{x}_N\}$ be the data set, assume that for any real feature vector $\mathbf{x}_i$ its components $\{x_{ij}\}_j$ and its norm $\| \mathbf{x}_i \|$ are stored in a QRAM as floating point numbers, so the state $|\mathbf{x}_i\rangle = \| \mathbf{x}_i \|^{-1} \sum_j x_{ij} |j\rangle$ can be retrieved in $O(\log d)$ steps. Assuming the centroids are classically calculated and then encoded into quantum states, we can call **Qdist** to calculate the Euclidean distance between data vectors and centroids. Since a single iteration of K-means requires NK distance calculations then the complexity per iteration of the **Qdist**-based algorithm is $O(NK \log d)$.

An improved version of quantum K-means is based on the adiabatic theorem for returning the following quantum state in time $O(K \log(NKd))$ [Ll13]:

$$|\Psi\rangle = \frac{1}{\sqrt{N}} \sum_{P,i:\mathbf{x}_i \in P} |P\rangle|i\rangle, \tag{6.2.3}$$

where the sum $\sum_P$ is taken over the K clusters. In the quantum superposition (6.2.3), the states $|i\rangle$ which encode the indexes of the data instances $\mathbf{x}_i$ are correlated to the states $|P\rangle$ encoding the corresponding cluster P. The clustering structure is found sampling the quantum output (6.2.3). Before the description of the preparation of the state (6.2.3), let us generalize the **Qdist** routine to avoid the classical calculation of the centroids. Consider a cluster of M datapoints $\{\mathbf{x}_1, ..., \mathbf{x}_M\}$ and an arbitrary data instance $\mathbf{x}$. This distance calculation between $\mathbf{x}$ and the centroid of the considered cluster requires the $\log d$-qubit register (with Hilbert space H_d) to store the d-dimensional vectors within the amplitude encoding and a $\log M$-qubit register (with Hilbert space H_M) to store the indexes of the data instances as states of the computational basis $\{|m\rangle\}_{m=1,...,M}$. The procedure is based on the state:

$$|\psi\rangle = \frac{1}{\sqrt{2}} \left(|0\rangle|\mathbf{x}\rangle + \frac{1}{\sqrt{M}} \sum_{m=1}^{M} |m\rangle|\mathbf{x}_m\rangle \right) \in \mathsf{H}_M \otimes \mathsf{H}_d, \tag{6.2.4}$$

which can be retrieved from the QRAM in time $O(\log Md)$, and the state:

$$|\varphi\rangle = \frac{1}{\sqrt{Z}} \left(\parallel \mathbf{x} \parallel |0\rangle - \frac{1}{\sqrt{M}} \sum_{m=1}^{M} \parallel \mathbf{x}_m \parallel |m\rangle \right) \in \mathsf{H}_M, \qquad (6.2.5)$$

where $Z = \parallel \mathbf{x} \parallel^2 + \frac{1}{M} \sum_m \parallel \mathbf{x}_m \parallel^2$, which can be constructed as follows. Using the quantum access to the norms of the vectors, we can simulate the Hamiltonian:

$$\mathcal{H} = \left(\parallel \mathbf{x} \parallel |0\rangle\langle 0| + \sum_m \parallel \mathbf{x}_m \parallel |m\rangle\langle m| \right) \otimes \sigma_x, \qquad (6.2.6)$$

to evolve the state (where there is an ancillary qubit):

$$|\Psi_0\rangle = \frac{1}{\sqrt{2}} \left(|0\rangle - \frac{1}{M} \sum_{m=1}^{M} |m\rangle \right) |0\rangle \in \mathsf{H}_M \otimes \mathsf{H}_{ancilla}, \qquad (6.2.7)$$

into the state:

$$|\Psi_1\rangle = e^{-i\mathcal{H}t}|\Psi_0\rangle = \frac{1}{\sqrt{2}} \left(\cos(\parallel \mathbf{x} \parallel t)|0\rangle - \frac{1}{\sqrt{M}} \sum_{m=1}^{M} \cos(\parallel \mathbf{x}_m \parallel t)|m\rangle \right) |0\rangle -$$

$$- \frac{i}{\sqrt{2}} \left(\sin(\parallel \mathbf{x} \parallel t)|0\rangle - \frac{1}{\sqrt{M}} \sum_{m=1}^{M} \sin(\parallel \mathbf{x}_m \parallel t)|m\rangle \right) |1\rangle.$$

Assuming to set the time interval such that $\parallel \mathbf{x} \parallel t, \parallel \mathbf{x}_m \parallel t \ll 1$, we have:

$$|\Psi_1\rangle \simeq \frac{1}{\sqrt{2}} \left(|0\rangle - \frac{1}{\sqrt{M}} \sum_{m=1}^{M} |m\rangle \right) |0\rangle - \frac{i}{\sqrt{2}} \left(\parallel \mathbf{x} \parallel t|0\rangle - \frac{1}{\sqrt{M}} \sum_{m=1}^{M} \parallel \mathbf{x}_m \parallel t|m\rangle \right) |1\rangle.$$

Measuring the ancillary qubit, the probability of the outcome 1 is:

$$\mathbb{P}_{ancilla}(1) = \frac{t^2}{2} \left(\parallel \mathbf{x} \parallel^2 + \frac{1}{M} \sum_{m=1}^{M} \parallel \mathbf{x}_m \parallel^2 \right) = \frac{t^2}{2} Z. \qquad (6.2.8)$$

The post-measurement state is $|\varphi\rangle$ and, repeating the procedure, we can estimate Z. Once prepared the states (6.2.4) and (6.2.5), we perform the SWAP test over the two $\log M$-qubit registers. In a similar way to what we did in Section 5.3, we compute the inner product between $|\varphi\rangle$ and the tensor factor of $|\psi\rangle$ in H_M obtaining the non-normalized vector:

$$\Phi = \frac{1}{\sqrt{2Z}} \left(\parallel \mathbf{x} \parallel |\mathbf{x}\rangle - \frac{1}{\sqrt{M}} \sum_{m=1}^{M} \parallel \mathbf{x}_m \parallel |\mathbf{x}_m\rangle \right). \qquad (6.2.9)$$

The length of Φ can be estimated by the success probability of the SWAP test providing the distance between $\mathbf{x}$ and the centroid of the cluster $\{\mathbf{x}_1, ..., \mathbf{x}_M\}$:

$$\parallel \Phi \parallel^2 = \frac{1}{2Z}\left\|\mathbf{x} - \frac{1}{M}\sum_{m=1}^{M}\mathbf{x}_m\right\|^2. \tag{6.2.10}$$

The first step of the preparation of the clustering state (6.2.3) is the selection of K vectors, indexed by i_P, as initial seeds for each cluster. The initial quantum state is:

$$\frac{1}{\sqrt{KN}}\sum_{P',i=1,...,N}|P'\rangle|i\rangle\left(\frac{1}{\sqrt{K}}\sum_{P}|P\rangle|i_P\rangle\right)^{\otimes q}, \tag{6.2.11}$$

the multiple copies of the seed state $\frac{1}{\sqrt{K}}\sum_P|P\rangle|i_\mathbf{c}\rangle$ allows to estimate the distance $\|\mathbf{x}_i - \mathbf{x}_{i_{P'}}\|^2$ in the $P'i$ component of the state (6.2.11) with the ability of simulating the Hamiltonian:

$$\mathcal{H}_F = \sum_{P',i=1,...,N}\|\mathbf{x}_i - \mathbf{x}_{i_{P'}}\|^2|P'\rangle\langle P'|\otimes|i\rangle\langle i|. \tag{6.2.12}$$

The initial Hamiltonian $\mathcal{H}_I := \mathbb{I} - |\phi\rangle\langle\phi|$ where $|\phi\rangle := \frac{1}{\sqrt{KN}}\sum_{P',i}|P'\rangle|i\rangle$ can be adiabatically deformed into $\mathcal{H}_F$ whose ground state is:

$$|\phi_1\rangle = \frac{1}{\sqrt{N}}\sum_{P,i:\mathbf{x}_i\in P}|P\rangle|i\rangle, \tag{6.2.13}$$

where $|i\rangle$ is correlated to $|P\rangle$ with the closest seed vector i_P. Measuring the "cluster register", one can obtain the superposition of the elements of any cluster:

$$|\phi_P\rangle = \frac{1}{|P|}\sum_{i:\mathbf{x}_i\in P}|i\rangle, \tag{6.2.14}$$

so the number of elements $|P|$ of each cluster P can be estimated constructing multiple copies of $|\phi_1\rangle$.

For the re-clustering step, let us assume to construct q copies of the state $|\phi_1\rangle$. By constructing the cluster state (6.2.14) and by the quantum estimation of the distance:

$$\|\mathbf{x}_i - \mathbf{c}_P\|^2 = \left\|\mathbf{x}_i - \frac{1}{|P|}\sum_{\mathbf{x}\in P}\mathbf{x}\right\|^2, \tag{6.2.15}$$

we can simulate the following Hamiltonian:

$$\mathcal{H}_{F'} = \sum_{P',i=1,\dots,N} \|\mathbf{x}_i - \mathbf{c}_{P'}\|^2 |P'\rangle\langle P'| \otimes |i\rangle\langle i| \otimes \mathbb{I}^{\otimes q}. \tag{6.2.16}$$

Now we can implement an adiabatic evolution from the initial state:

$$|\phi\rangle|\phi_1\rangle^{\otimes q} = \frac{1}{\sqrt{NK}} \sum_{P',i=1,\dots,N} |P'\rangle|i\rangle|\phi_1\rangle^{\otimes q}, \tag{6.2.17}$$

towards the ground state of the Hamiltonian $\mathcal{H}_{F'}$:

$$|\phi_2\rangle|\phi_1\rangle^{\otimes q} = \left(\frac{1}{\sqrt{N}} \sum_{P',i:\mathbf{x}_i \in P} |P'\rangle|i\rangle \right) |\phi_1\rangle^{\otimes q}, \tag{6.2.18}$$

where each state $|i\rangle$ is correlated with the state $|P'\rangle$ encoding the cluster with the closest centroid. Once created q copies of $|\phi_2\rangle$ a new iteration can be performed. Since the convergence of the Lloyd's algorithm is fast, a small number of steps is required to achieve the state (6.2.3). In order to estimate the complexity of one iteration, let us observe that the calculation of the distances $\|\mathbf{x}_i - \mathbf{c}_P\|^2$ in parallel requires $O(\log KNd)$ steps for state retrieval from the QRAM. Since $|\langle\phi|\phi_2\rangle|^2 = O(1/K)$, the execution of the adiabatic evolution from the state (6.2.17) to the state (6.2.18) takes time $O(K)$. Therefore, the complexity of a single iteration in the quantum K-means based on the sampling over the quantum output (6.2.3) is $O(K \log(KNd))$ that improves the performance of the quantum K-means based only on the calls to Qdist which has complexity $O(KN \log d)$.

6.3   Quantum K-medians

The K-medians algorithm is similar to K-means but the centroids are calculated with the medians in place of the means. This approach presents the advantage to generate centroids that belongs to the dataset, in fact the centroids generated with the K-means may lie outside the manifold where data points are located. The main disadvantage of K-medians is that it may do not converge.

Let $\{\mathbf{x}_1, \cdots, \mathbf{x}_N\}$ be a dataset described in a d-dimensional feature space and $P = P_1, \dots, P_K$ be a partition of $\{\mathbf{x}_1, \cdots, \mathbf{x}_N\}$ into K clusters. The centroid of P_j is defined as the median of its elements:

$$\mathbf{c}_j := \operatorname*{argmin}_{\mathbf{y} \in P_j} \sum_{\mathbf{x} \in P_j} \| \mathbf{x} - \mathbf{y} \| \equiv \operatorname*{argmin}_{\mathbf{y} \in P_j} F_j(\mathbf{y}). \tag{6.3.1}$$

We know that the Euclidean distance between d-dimensional vectors can be calculated in time $O(\log d)$ calling Qdist assuming the QRAM data structure. Here there is the distance calculation between each data point and each centroid like in K-means but also the distance calculation between data points for the calculation of the median. To compute the function $F_j(\mathbf{y}) = \sum_{\mathbf{x} \in P_j} \| \mathbf{x} - \mathbf{y} \|$, we can repeatedly call the routine Qdist. Then we call the Dürr-Høyer algorithm to efficiently minimize F_j obtaining the median $\mathbf{c}_j$. In Algorithm 4, there is the quantum K-medians clustering based on repeated calls to the subroutines Qdist and **Dürr-Høyer**.

The classical complexity of the median computation is quadratic in the number of considered points and linear in the dimension of the space. The average cardinality of a cluster is N/K then the complexity per iteration of the classical K-medians is $O(\frac{N^2}{K}d)$. In the quantum version, calling Qdist as an oracle and running the Dürr-Høyer algorithm, a centroid can be calculated in time $O\left(\frac{N}{K}\sqrt{\frac{N}{K}}\log d\right)$ then the complexity per iteration of the quantum K-medians is $O\left(\frac{N^{\frac{3}{2}}}{\sqrt{K}}\log d\right)$.

Input: Data set $\{\mathbf{x}_1, \cdots, \mathbf{x}_N\}$, number of clusters K
Result: Partition of $\{\mathbf{x}_1, \cdots, \mathbf{x}_N\}$ into K clusters

1 initialize K centroids $\mathbf{c}_1, ..., \mathbf{c}_K$ from the elements of the dataset V;
2 **repeat**
3   **foreach** $i \leftarrow 1, ..., N$ **do**
4   Qdist $(\mathbf{x}_i, \mathbf{c}_j)$ $\forall j = 1, ..., K$ and find the nearest centroid to $\mathbf{x}_i$;
5   **end**
6   **foreach** $j \leftarrow 1, ..., K$ **do**
7   construct the cluster $P_j = \{\mathbf{x}_i : \mathbf{c}_j$ is the nearest centroid$\}$;
8   use Qdist and **Dürr-Høyer** to compute and minimize F_j;
9   **end**
10 **until** *convergence*;
11 **return** $P_1, ..., P_K$

Algorithm 4: *Quantum K-medians algorithm*

6.4 Quantum divisive clustering

The divisive clustering can be summarized as follows: initially, each data point belongs to a single cluster. Then the cluster is split into two subclusters, the division is done choosing the two data points that are farthest apart as initial seeds, so the other points are attached to their closest seed. The division is recursively applied to the subclusters. For N data points, finding the two farthest points in the dataset classically requires $\Theta(N^2)$ comparisons, furthermore the classical calculation of Euclidean distance in $\mathbb{R}^d$ takes time $O(d)$.

Input: Data set $\{\mathbf{x}_1, \cdots, \mathbf{x}_N\}$
Result: Farthest data points $\mathbf{x}_i$, $\mathbf{x}_j$
1 randomly select two indexes i and j;
2 $d_{max} \leftarrow \| \mathbf{x}_i - \mathbf{x}_j \|$;
3 **repeat**
4   **foreach** $i \leftarrow 1, ..., N$ **do**
5        $\| \mathbf{x}_i - \mathbf{x}_j \| \leftarrow \mathsf{Qdist}(\mathbf{x}_i, \mathbf{x}_j)$;
6   **end**
7   apply Grover's algorithm to find i, j such that $\| \mathbf{x}_i - \mathbf{x}_j \| > d_{max}$;
8     $d_{max} \leftarrow \| \mathbf{x}_i - \mathbf{x}_j \|$;
9 **until** *no new i, j are found*;
10 **return** $(\mathbf{x}_i, \mathbf{x}_j)$

Algorithm 5: *Quantum search of the two data points farthest from each other.*

Input: Cluster $\mathcal{C}$
Result: Clusters in $\mathcal{C}$
1 **if** *points of $\mathcal{C}$ are sufficiently similar* **then**
2   **return** $\mathcal{C}$
3 **else**
4     $(\mathbf{x}_i, \mathbf{x}_j) \leftarrow Alg5(\mathcal{C})$;
5   **foreach** $\mathbf{x} \in \mathcal{C}$ **do**
6   attach $\mathbf{x}$ to the closest between $\mathbf{x}_i$ and $\mathbf{x}_j$;
7   **end**
8   construct $\mathcal{C}_i = \{\mathbf{x} \in \mathcal{C}$ attached to $\mathbf{x}_i\}$;
9   construct $\mathcal{C}_j = \{\mathbf{x} \in \mathcal{C}$ attached to $\mathbf{x}_j\}$;
10   recurvsive call on $\mathcal{C}_i$ and $\mathcal{C}_j$;
11 **end**

Algorithm 6: *Quantum divisive clustering.*

With a quantum machine, one can calculate the Eiclidean distance with exponential speedup as done in quantum K-means and K-medians and use an iterative procedure with repeated calls to Grover's algorithm to find the maximum distance within the cluster. Algorithm 5 based on Grover finds the two farthest points from each other in a data set of N elements in time $\Theta(N)$. The quantum divisive clustering algorithm, as proposed in [Aï07], is outlined in Algorithm 6. Assuming that Algorithm 6 splits a cluster into balanced subclusters the running time $T(N)$ is given by the asymptotic recurrence $T(N) = 2T(N/2) + \Theta(N)$ that is $\Theta(N \log N)$ by the *master theorem* of the algorithm analysis [Co09].

6.5 Clustering with a quantum annealer

In the previous sections we have discussed different methods to cluster a dataset with a universal quantum computer. Nevertheless, also a quantum annealer can be used for clustering [Ba18]. In particular, a problem of 2-means clustering can be formulated in terms of a QUBO problem that can be treated by a quantum annealer. Assume that our task is clustering a dataset of N real feature vectors of dimension d into two clusters P_1 and P_2. The 2-means clustering is given by the following optimization problem:

$$\underset{P_1, P_2}{\operatorname{argmin}} \sum_{i=1,2} \sum_{\mathbf{x} \in P_i} \| \mathbf{x} - \mathbf{c}_i \|^2, \tag{6.5.1}$$

where $\mathbf{c}_i$ is the centroid of the cluster P_i calculated as the mean of its elements. The problem (6.5.1) can be reformulated as:

$$\underset{P_1, P_2}{\operatorname{argmax}} \sum_{i,j=1,2} N_i N_j \| \mathbf{c}_i - \mathbf{c}_j \|^2, \tag{6.5.2}$$

where N_i is the number of data points in the cluster P_i. The problem (6.5.2) is an integer programming problem that is NP-hard. Assuming that the size of the two cluster is comparable, that is $N_1 \simeq N_2 \simeq N/2$, the objective function to maximize can be approximated as follows:

$$F(P_1, P_2) = 2 N_1 N_2 \| \mathbf{c}_1 - \mathbf{c}_2 \|^2 \simeq 2 \| N_1 \mathbf{c}_1 - N_2 \mathbf{c}_2 \|^2 . \tag{6.5.3}$$

In order to use a quantum annealer to cluster the dataset, we need to cast the problem:

$$\underset{P_1, P_2}{\operatorname{argmax}} \| N_1 \mathbf{c}_1 - N_2 \mathbf{c}_2 \|^2, \tag{6.5.4}$$

into an Ising model. Let us define the binary indicator vectors $\mathbf{s}_1, \mathbf{s}_2 \in \{0,1\}^N$ so that $s_{kj} = 1$ if and only if $\mathbf{x}_j \in P_k$ and $s_{kj} = 0$ otherwise for $k = 1, 2$ and $j = 1, ..., N$. Therefore:

$$N_i \mathbf{c}_i = X^T \mathbf{s}_i \qquad i = 1, 2, \tag{6.5.5}$$

where X is the data matrix where the rows are given by the data instances and the columns correspond to the features. The problem (6.5.4) can be now formulated in terms of an optimization over the indicator vectors and re-written in the following form:

$$\underset{\mathbf{z} \in \{-1,+1\}^N}{\operatorname{argmin}} \left\{ - \| X^T \mathbf{z} \|^2 \right\} = \underset{\mathbf{z} \in \{-1,+1\}^N}{\operatorname{argmin}} \mathbf{z}^T Q \mathbf{z}, \tag{6.5.6}$$

where $\mathbf{z} := \mathbf{s}_1 - \mathbf{s}_2$ and $Q := -XX^T$. The obtained formulation of the problem presents an ambiguity: if $\mathbf{z}^*$ is a solution then $-\mathbf{z}^*$ is so. Thus we can remove a degree of freedom from the model fixing the last component of the vector $\mathbf{z}$ to 1. Under the choice $z_N = 1$, the considered QUBO problem in the Ising form is:

$$\underset{\mathbf{z}\in\{-1,1\}^{N-1}}{\arg\min} \left[\sum_{i,j=1}^{N-1} Q_{ij}z_iz_j + 2\sum_{i=1}^{N-1} Q_{Ni}z_i\right] = \underset{\mathbf{z}\in\{-1,1\}^{N-1}}{\arg\min} \left[\sum_{i<j} \theta_{ij}z_iz_j + \sum_j \theta_j z_j\right] (6.5.7)$$

The entries of the solution vector $\mathbf{z}^*$ correspond to the cluster assignment of each data instance: if the ith entry of $\mathbf{z}^*$ is $+1$ then the data instance $\mathbf{x}_i$ is assigned to the cluster of $\mathbf{x}_N$ otherwise it is assigned to the other cluster. In order to solve the problem (6.5.7), the weights of the quantum annealers must be initialized as: $\theta_i = 2\,\mathbf{x}_N\cdot\mathbf{x}_i$ and $\theta_{ij} = \mathbf{x}_i\cdot\mathbf{x}_j$ for $i,j = 1, ..., N-1$. The data instance $\mathbf{x}_N$ corresponds to the removed degree of freedom and is used to flag one of the two clusters. In the existing quantum annealers, the topology of the architecture is highly sparse then many qubit couplings are set to zero, however in the model the coupling terms are zero if and only if the corresponding data instances are orthogonal in the feature space. Therefore a minor embedding on the D-Wave machine [DW21b] or other techniques of non-trivial problem encoding [PB19] are required.

Chapter 7

Quantum classification

In supervised learning, a general classification problem is defined as the assignment of labels to new data instances given a training set of already labelled (classified) data. In this section we introduce some quantum algorithms to make predictions on labels of previously unseen data instances: two examples of quantum distance-based classifiers, a quantum versions of the k-nearest neighbors algorithm, and the quantum support vector machine. Moreover, we overview a quantum-inspired classifier that is an algorithm based on the quantum formalism but devised for classical machines.

7.1 Distance-based quantum classification

The general strategy of distance-based classification is calculating the distance in the feature space of a new data instance with respect to the training data. The new input can be classified evaluating the labels of the elements in the training set according to their distance from the test point. Therefore if we have N data points described in a d-dimensional feature space, the classical calculation of the distances between a new datum and all the training data takes time $O(Nd)$. In particular, binary classification can be formalized as follows: Let $\{(\mathbf{x}_1, y_1), ..., (\mathbf{x}_N, y_N)\}$ be a training set of data points $\mathbf{x}_i \in \mathbb{R}^d$ with the corresponding labels $y_i \in \{-1, 1\}$. Given a test data point $\mathbf{x} \in \mathbb{R}^d$, the goal is to find the corresponding label $y \in \{-1, 1\}$. Let us consider the quantum implementation of a binary classifier, based on the proposal [Sc17], that assigns the new label to $\mathbf{x}$ by evaluating:

$$y = \text{sgn}\left[\sum_{i=1}^{N} y_i \left(1 - \frac{1}{4} \| \mathbf{x} - \mathbf{x}_i \|^2\right)\right]. \tag{7.1.1}$$

© The Author(s), under exclusive license to Springer Nature Singapore Pte Ltd. 2023
D. Pastorello, *Concise Guide to Quantum Machine Learning*,
Machine Learning: Foundations, Methodologies, and Applications,
https://doi.org/10.1007/978-981-19-6897-6_7

This is a model where the labels of all training points contribute to the prediction and the relevance of the contribution decreases with the square distance. Let us assume the normalization of the data set, so $\mathbf{x}_i^T \mathbf{x}_i = 1$ for any $i = 1, ..., N$ and the standardization to zero mean and unit variance. By amplitude encoding, $\mathbf{x}_i$ is represented by the following quantum state in a d-dimensional Hilbert space H_d:

$$|\psi_{\mathbf{x}_i}\rangle = \sum_{j=1}^{d} x_{ij}|j\rangle, \tag{7.1.2}$$

where $\{|j\rangle\}_{j=1,...,d}$ is the computational basis of H_d. Moreover, we consider a N-level quantum system, described in the Hilbert space H_N, as a register to keep track of the i-th training point within the basis encoding. Using a suitable state preparation scheme or a retrieval from a QRAM, a composite system is prepared in the state:

$$|\Psi_0\rangle = \frac{1}{\sqrt{2N}} \sum_{i=1}^{N} |i\rangle(|0\rangle|\psi_{\mathbf{x}}\rangle + |1\rangle|\psi_{\mathbf{x}_i}\rangle)|y_i\rangle \in \mathsf{H}^{(N)} \otimes \mathsf{H}_a \otimes \mathsf{H}^{(d)} \otimes \mathsf{H}_b, \tag{7.1.3}$$

where $\{|i\rangle\}_{i=1,...,N}$ is the computational basis of H_N used to flag any training point, there is an ancillary qubit described in H_a that is entangled with the register encoding the data points by amplitudes and H_b is another 1-qubit register encoding the label of any training point. By the application of a Hadamard gate on H_a we get:

$$|\Psi_1\rangle = \frac{1}{2\sqrt{N}} \sum_{i=1}^{N} |i\rangle[|0\rangle(|\psi_{\mathbf{x}}\rangle + |\psi_{\mathbf{x}_i}\rangle) + |1\rangle(|\psi_{\mathbf{x}}\rangle - |\psi_{\mathbf{x}_i}\rangle)]|y_i\rangle. \tag{7.1.4}$$

Now let us assume to perform a measurement on the qubit of the register H_a, if the outcome is 0 then the post-measurement state:

$$|\Psi_2\rangle = \frac{1}{2\sqrt{N\mathbb{P}(0)}} \sum_{i=1}^{N} |i\rangle|0\rangle(|\psi_{\mathbf{x}}\rangle + |\psi_{\mathbf{x}_i}\rangle)|y_i\rangle$$

$$= \frac{1}{2\sqrt{N\mathbb{P}(0)}} \sum_{i=1}^{N} \sum_{j=1}^{d} |i\rangle|0\rangle(x_j + x_{ij})|j\rangle|y_i\rangle, \tag{7.1.5}$$

where $\mathbb{P}(0)$ is the probability of obtaining the outcome 0:

$$\mathbb{P}(0) = \frac{1}{4N} \sum_{i=1}^{N} \| \mathbf{x} + \mathbf{x}_i \|^2, \tag{7.1.6}$$

if data are standardized to zero mean and unit variance then $\mathbb{P}(0) \simeq 0.5$. Now a measurement on the qubit of the register H_b gives the outcome 1 or -1 according to the probability:

$$\mathbb{P}_b(l) = \frac{1}{4Np_0} \sum_{i:y_i=l} \| \mathbf{x} + \mathbf{x}_i \|^2, \tag{7.1.7}$$

where $l = \pm 1$. Since the normalization to 1 of the data points implies that:

$$\frac{1}{4N} \sum_{i=1}^{N} \| \mathbf{x} + \mathbf{x}_i \|^2 = 1 - \sum_{i=1}^{N} \frac{1}{4N} \| \mathbf{x} - \mathbf{x}_i \|^2 = \frac{1}{N} \sum_{i=1}^{N} \left(1 - \frac{1}{4} \| \mathbf{x} - \mathbf{x}_i \|^2 \right) \tag{7.1.8}$$

we have that 7.1.7 is the probability to predict the class l for the new input consistently with the model 7.1.1. The result is probabilistic, then the algorithm must be repeated several times to obtain an accurate result. The number of measurements needed to estimate the correct classification to error ϵ grows as $O(\epsilon^{-2})$. Once prepared the initial state (7.1.3), the considered quantum classifier assigns a label to the new datum in constant time. The retrieval of $|\Psi_0\rangle$ from the QRAM requires $O(\log(Nd))$ operations, therefore the considered quantum classifier is exponentially faster than a classical execution of the same model.

Another quantum algorithm for distance-based binary classification that implements a model based on cosine similarity to predict the label of a new data instance is proposed in [PB21]. The main idea can be summarized as follows: The unclassified instance is put in the quantum superposition of the two possible classifications and then it is compared with all the training points with a single operation by quantum parallelism. If N is the number of training points and d is the dimension of the feature space where data are represented, this quantum algorithm classifies a new instance in time $O(\log(Nd))$ whereas the classical implementation of the same model presents a time complexity of $O(Nd)$. However, this proposal is not a quantum model that can be trained but an efficient quantum algorithm to implement the binary classification according to a considered cosine similarity-based model.

Let $X = \{\mathbf{x}_i, y_i\}_{i=0,...,N-1}$, with $\mathbf{x}_i \in \mathbb{R}^d$ and $y_i \in \{-1, 1\} \; \forall i \in \{0, ..., N-1\}$, be a training set of N data instances with two-valued labels that are represented in a real feature space of dimension d. Let $\mathbf{x} \in \mathbb{R}^d$ be a new data instance to be classified as either 1 or -1. The considered classification model for the quantum implementation is defined as follows:

$$y(\mathbf{x}) := \mathrm{sgn}\left(\sum_{i=0}^{N-1} y_i \cos(\mathbf{x}_i, \mathbf{x}) \right), \tag{7.1.9}$$

where *cosine similarity* is defined by:

$$\cos(\mathbf{x}, \mathbf{y}) := \frac{\mathbf{x} \cdot \mathbf{y}}{\| \mathbf{x} \| \| \mathbf{y} \|} \qquad \mathbf{x}, \mathbf{y} \in \mathbb{R}^d. \tag{7.1.10}$$

A typical example where cosine similarity is adopted for classification and clustering is text analysis [Ho12, Al16, FA17]. Furthermore, in the case of normalized data vectors, (7.1.10) reduces to the dot product and it is directly related to the Euclidean distance by $\| \mathbf{x} - \mathbf{y} \| = \sqrt{2(1 - \mathbf{x} \cdot \mathbf{y})}$. In the model (7.1.9), any training vector contributes to the prediction of the new label and such a contribution is weighted by the cosine similarity with the new instance.

On one hand, the classical calculation of the new label according to (7.1.9) requires N cosine similarities each computed in time $O(d)$ then the overall time complexity is $O(Nd)$. On the other hand, the quantum implementation of the model (7.1.9) introduced in the following is based on the encoding of data vectors into amplitudes of a coherent superposition of quantum states (amplitude encoding), on a suitable state preparation, and on the so-called SWAP test [Bu01]. Data vectors $\mathbf{x}_i \in \mathbb{R}^d$ are stored in a n-qubit register within the amplitude encoding. There is also a $\log N$-qubit register, with Hilbert space $\mathsf{H}_{index} \simeq (\mathbb{C}^2)^{\otimes \log N}$, to encode the indexes of training data vectors and construct the state:

$$|X\rangle = \frac{1}{\sqrt{N}} \sum_{i=0}^{N-1} |i\rangle |\mathbf{x}_i\rangle |b_i\rangle \in \mathsf{H}_{index} \otimes \mathsf{H}_n \otimes \mathsf{H}_l, \tag{7.1.11}$$

where H_l is the Hilbert space of a single qubit used for encoding the values of the labels with $b_i = \frac{1 - y_i}{2} \in \{0, 1\}$, then $|b_i\rangle$ is eigenstate of the Pauli matrix σ_z with eigenvalue y_i. The entangled state (7.1.11) encodes the training set X as a coherent superposition of its elements with respective labels, note that just one qubit is sufficient for the encoding of all the labels. Moreover, in the same registers construct the state:

$$|\psi_{\mathbf{x}}\rangle = \frac{1}{\sqrt{N}} \sum_{i=0}^{N-1} |i\rangle |\mathbf{x}\rangle |-\rangle \in \mathsf{H}_{index} \otimes \mathsf{H}_n \otimes \mathsf{H}_l, \tag{7.1.12}$$

where the label qubit is in the state $|-\rangle = \frac{1}{\sqrt{2}}(|0\rangle - |1\rangle)$, so the new data vector $\mathbf{x}$ is represented in a quantum superposition of the two possible classifications. Now consider an ancillary qubit, called qubit a, and prepare the state:

$$\frac{1}{\sqrt{2}} \left(|X\rangle |0\rangle + |\psi_{\mathbf{x}}\rangle |1\rangle \right) \in \mathsf{H}_{index} \otimes \mathsf{H}_n \otimes \mathsf{H}_l \otimes \mathsf{H}_a, \tag{7.1.13}$$

that can be retrieved from the QRAM in time $O(\log(Nd))$. Now perform the SWAP test between a second ancillary qubit, called qubit b, prepared in $|+\rangle = \frac{1}{\sqrt{2}}(|0\rangle + |1\rangle)$ and the qubit a, moreover consider another qubit, say c, prepared in $|0\rangle$ to control the Fredkin gate. A straightforward calculation shows that the probability to obtain the outcome 1 measuring the qubit c is:

$$\mathbb{P}(1) = \frac{1}{4}(1 - \langle X | \psi_{\mathbf{x}}\rangle), \tag{7.1.14}$$

that is directly related to (7.1.9), in fact:

$$\langle X | \psi_{\mathbf{x}}\rangle = \frac{1}{N} \sum_{i,k=0}^{N-1} \langle i|k\rangle \langle \mathbf{x}_i|\mathbf{x}\rangle \langle b_i|-\rangle \tag{7.1.15}$$

$$= \frac{1}{N\sqrt{2}} \sum_{i=0}^{N-1} \langle \mathbf{x}_i|\mathbf{x}\rangle (\langle b_i|0\rangle - \langle b_i|1\rangle)$$

$$= \frac{1}{N\sqrt{2}} \sum_{i=0}^{N-1} y_i \cos(\mathbf{x}_i, \mathbf{x}),$$

once applied the identities $\langle i|k\rangle = \delta_{ik}$ and $\langle b_i|0\rangle - \langle b_i|1\rangle = 1 - 2b_i = y_i$ for any $i = 0, ..., N-1$. Therefore, the probability $\mathbb{P}(1)$ is related to the prediction of the label of $\mathbf{x}$, according to the model (7.1.9), by means of:

$$y(\mathbf{x}) = \text{sgn}\left[1 - 4\mathbb{P}(1)\right]. \tag{7.1.16}$$

After the retrieval of the state (7.1.13) from a QRAM in time $O(\log(Nd))$, an appropriate SWAP test is sufficient to classify the new instance $\mathbf{x}$ according to the model (7.1.9) with an exponential speedup with respect to the classical calculation. However, the procedure *preparation+test* must be repeated several times for sampling the qubit c to estimate $\mathbb{P}(1)$ as the success probability of a Bernoulli trial, so an estimation within an error ϵ requires a number of repetitions growing as $O(\epsilon^{-2})$ as provided by the binomial proportion confidence interval. Thus the overall time complexity of Algorithm 7 is $O(\epsilon^{-2} \log (Nd))$.

One can provide an example of implementation of the introduced quantum binary classifier for $N = 2$ and $d = 2$. Consider a training set of two-dimensional data instances given by $X = \{(\mathbf{x}_0, y_0), (\mathbf{x}_1, y_1)\}$ where $\mathbf{x}_0 = (1,0)$, $y_0 = 1$ and $\mathbf{x}_1 = (0.718, 0.696)$, $y_1 = -1$. Let $\mathbf{x} = (0.884, 0.468)$ be the unlabelled data instance to classify. In this simple example, with normalized data vectors and only two training points, the model (7.1.9) predicts the label of $\mathbf{x}$ as a *nearest neighbor* then it returns $y = -1$.

Input: training set $X = \{\mathbf{x}_i, y_i\}_{i=0,...,N-1}$, unclassified instance $\mathbf{x}$.
Result: label y of $\mathbf{x}$.

1 **repeat**
2   initialize the register $\mathsf{H}_{index} \otimes \mathsf{H}_n \otimes \mathsf{H}_l$ and an ancillary qubit a in the state (7.1.13);
3   initialize a qubit b in the state $|-\rangle$;
4   perform the SWAP test on a and b with control qubit c prepared in $|0\rangle$;
5   measure qubit c;
6 **until** *desired accuracy on the estimation of* $\mathbb{P}(1)$;
7 Estimate $\mathbb{P}(1)$ as the relative frequency $\hat{\mathbb{P}}$ of outcome 1;
8 **if** $\hat{\mathbb{P}} > 0.25$ **then**
9   **return** $y = -1$
10 **else**
11   **return** $y = 1$
12 **end**

Algorithm 7: *Quantum implementation of the model (7.1.9).*

In order to run Algorithm 7, according to (7.1.11) and (7.1.12), we prepare the state:

$$|X\rangle = \frac{1}{\sqrt{2}}(|0\rangle|0\rangle|0\rangle + |1\rangle|\mathbf{x}_1\rangle|1\rangle), \tag{7.1.17}$$

and the state:

$$|\psi_{\mathbf{x}}\rangle = |+\rangle|\mathbf{x}\rangle|-\rangle. \tag{7.1.18}$$

The algorithm can be run on a IBM prototype quantum processor that is available within the cloud-based quantum computing service of *IBM Quantum* [IBM]. Consider the quantum circuit for the construction of the state (7.1.17):

$$\tag{7.1.19}$$

where $RY_{(0.49\pi)}$ is a rotation of 0.49π around the y-axis of the Bloch sphere mapping $|0\rangle$ into $|\mathbf{x}_1\rangle = 0.718|0\rangle + 0.696|1\rangle$. The circuit to construct the state (7.1.18) is simply:

$$\tag{7.1.20}$$

where the gate $RY_{(0.31\pi)}$ rotates $|0\rangle$ into $|\mathbf{x}\rangle = 0.884|0\rangle + 0.468|1\rangle$ and there is a bit-flip followed by a Hadamard gate to prepare the label qubit in $|-\rangle$.

In Fig. 7.1, there is the circuit, represented in the *IBM Quantum Composer*, implementing Algorithm 7 for the example just introduced. The qubits q_0, q_1, q_2 are the ancillas used for the SWAP test, q_3 is the 1-qubit index register, q_4 is an additional ancillary qubit necessary to implement the controls of the gate $RY_{(0.49\pi)}$, q_5 is the 1-qubit register for the amplitude encoding of data and q_6 is the qubit encoding the labels.

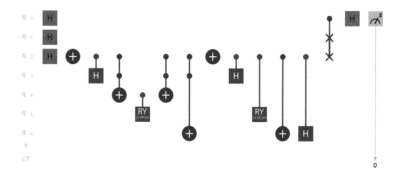

Fig. 7.1. Quantum circuit implementing lines 1-7 of Algorithm 1 in the IBM Quantum Composer where $N = 2$, $d = 2$, the training set is $\{((1,0),1), ((0.718, 0.696), -1)\}$ and the unclassified data vector is $\mathbf{x} = (0.884, 0.468)$.

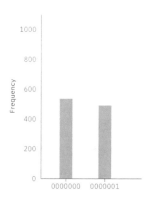

Fig. 7.2. Output statistic of the circuit reported in Fig.1 given by the IBM quantum processor *ibmq_16_melbourne*.

The performed run of the algorithm on the IBM quantum processor *ibmq_16_melbourne* provides 1024 shots for sampling the qubit q_0 (corresponding to the qubit c of the general algorithm description). The obtained statistic is shown in Fig. 6.2, the relative frequency of the outcome 1, used to estimate the probability $\mathbb{P}(1)$, is $\hat{\mathbb{P}} = 490/1024 \simeq 0.48$ then the label assigned to $\mathbf{x} = (0.884, 0.468)$ is $y = -1$ as expected.

Despite the correct classification in this test, a comparison with the result of the simulator *ibm_qasm_simulator* suggests that the considered quantum machine is too noisy for a good classification by means of Algorithm 7. The output statistic of the simulator provides $\hat{\mathbb{P}} = 273/1024 \simeq 0.27$ as the estimation of $\mathbb{P}(1)$. This result is consistent with the fact that the unclassified data vector $\mathbf{x}$ is close to the intermediate point between the training vectors. On the other hand, the classification performed by the real quantum machine, with a high value of $\hat{\mathbb{P}}$, appears more *rough*. Repeating the test with the same training points and the new unlabelled instance $\mathbf{x} = (0.951, 0.309)$ (whose correct classification is $y = 1$), the quantum machine fails. In fact, the quantum machine returns the relative frequency $\hat{\mathbb{P}} = 338/1024 \simeq 0.38$, so it classifies $\mathbf{x}$ as $y = -1$. On the same test, the simulator ibm_qasm simulator returns $\hat{\mathbb{P}} = 244/1024 \simeq 0.24$ classifying correctly. The observed lack of accuracy in classification depends on the low *quantum volume*[1] (QV = 8) of the considered quantum processor.

7.2   Quantum k-nearest neighbors

The classical *k-nearest neighbors* (kNN) is a supervised learning algorithm that classifies a new input $\mathbf{x}$ comparing the labels of the k data points in the training set that are closest to $\mathbf{x}$. The nearest neighbor algorithm is a special case of kNN with $k = 1$. Given a training set $\{(\mathbf{x}_i, \lambda_i)\}_{i=1,\dots,N}$ and a reference distance d in the feature space, the algorithm can be summarized by the following steps:

1. Compute $\mathsf{d}(\mathbf{x}, \mathbf{x}_i)$ for any $i = 1, \dots, N$.

2. Select the k training points that are nearest to $\mathbf{x}$.

3. Assign the label of the majority of the k neighbors to $\mathbf{x}$.

The complexity of the classification of a single data point is $O(Nd)$ where d is the dimension of data vectors. There is no a general strategy to choose k a priori and usually an hyperparameter tuning is done [Zh17].

Quantum implementations of kNN based on the Hamming distance have been proposed [Sc14, Ru18] and are inspired by the structure of the quantum associative memory [Tr01] that we describe in Section 8.1. A quantum nearest neighbor

---

[1]Quantum volume has been defined in [Mo18] as a metric to measure the capability and robustness of a quantum computer. However, in this book we do not explicitly deal with this figure of merit.

algorithm in terms of Euclidean distance based on amplitude amplification and Dürr-Høyer algorithm has been proposed in [Wi15]. In this section we describe another quantum kNN algorithm [Ba20, Ma21] evaluating the square cosine similarity to select the neighbors:

$$| \cos(\mathbf{x}, \mathbf{y})|^2 = \left| \frac{\mathbf{x} \cdot \mathbf{y}}{\| \mathbf{x} \| \, \| \mathbf{y} \|} \right|^2 . \tag{7.2.1}$$

Given the training set $X = \{\mathbf{x}_i, y_i\}_{i=0,...,N-1}$ and the unclassified instance $\mathbf{x}$. The selection of the k nearest neighbors can be done implementing the SWAP test to estimate the overlaps $|\langle \mathbf{x}|\mathbf{x}_i \rangle|^2$. Consider two n-qubit register (two copies of the main register where data are encoded within the amplitude encoding assuming $d = 2^n$), the index register for tracking the training points, and an ancillary qubit prepared in the overall state:

$$\frac{1}{\sqrt{N}} \sum_{i=0}^{N-1} |i\rangle |\mathbf{x}_i\rangle |\mathbf{x}\rangle |0\rangle \in \mathsf{H}_{index} \otimes \mathsf{H}_n \otimes \mathsf{H}_n \otimes \mathsf{H}_a. \tag{7.2.2}$$

The state (7.2.2) requires $O(\log(Nd))$ steps to be retrieved from the QRAM and provides the superposition of the training points and the new data instance stored in two different registers. Now perform the SWAP test on the two n-qubit registers controlled by the ancillary qubit obtaining:

$$|\Psi\rangle = \frac{1}{2\sqrt{N}} \sum_{i=0}^{N-1} |i\rangle \left[(|\mathbf{x}_i\rangle|\mathbf{x}\rangle + |\mathbf{x}\rangle|\mathbf{x}_i\rangle)|0\rangle + (|\mathbf{x}_i\rangle|\mathbf{x}\rangle - |\mathbf{x}\rangle|\mathbf{x}_i\rangle)|1\rangle \right]. \tag{7.2.3}$$

The probability to obtain the outcome 0 by a measurement process on the ancillary qubit in the state (7.2.3) is:

$$\mathbb{P}(0) = \langle \Psi | P_0 \Psi \rangle \quad \text{with} \quad P_0 = \mathbb{I}_{index,n} \otimes |0\rangle\langle 0|, \tag{7.2.4}$$

where $\mathbb{I}_{index,n}$ is the identity operator on the Hilbert space $\mathsf{H}_{index} \otimes \mathsf{H}_n^{\otimes 2}$ and $|0\rangle\langle 0|$ is the projector onto the 1-dimensional subspace of H_a spanned by $|0\rangle$, as provided by the postulates of quantum mechanics. Therefore:

$$\mathbb{P}(0) = \frac{1}{4N} \sum_{i,l=0}^{N-1} \langle i|l\rangle (\langle \mathbf{x}_i|\langle \mathbf{x}| + \langle \mathbf{x}|\langle \mathbf{x}_i|)(|\mathbf{x}_l\rangle|\mathbf{x}\rangle + |\mathbf{x}\rangle|\mathbf{x}_l\rangle) \tag{7.2.5}$$

$$= \frac{1}{4N} \sum_{i=0}^{N-1} (2 + 2\langle \mathbf{x}_i|\mathbf{x}\rangle\langle \mathbf{x}|\mathbf{x}_i\rangle) = \frac{1}{2} + \frac{1}{2N} \sum_{i=0}^{N-1} |\langle \mathbf{x}|\mathbf{x}_i\rangle|^2.$$

A similar calculation, considering the projector $P_1 = \mathbb{I} \otimes |1\rangle\langle 1|$, yields $\mathbb{P}(1) = \langle \Psi | P_1 \Psi \rangle = \frac{1}{2} - \frac{1}{2N} \sum_{i=0}^{N-1} |\langle \mathbf{x}_i | \mathbf{x} \rangle|^2$. Thus the probability to obtain outcome $\alpha \in \{0, 1\}$ by a measurement process on the ancillary qubit is:

$$\mathbb{P}(\alpha) = \frac{1}{2} + (-1)^\alpha \frac{1}{2N} \sum_{i=0}^{N-1} |\langle \mathbf{x} | \mathbf{x}_i \rangle|^2, \tag{7.2.6}$$

and the corresponding post-measurement state of the registers is:

$$|\Psi_\alpha\rangle = \frac{\sum_{i=0}^{N-1} |i\rangle (|\mathbf{x}_i\rangle |\mathbf{x}\rangle + (-1)^\alpha |\mathbf{x}\rangle |\mathbf{x}_i\rangle)}{\sqrt{2 \left(N + (-1)^\alpha \sum_{i=0}^{N-1} |\langle \mathbf{x} | \mathbf{x}_i \rangle|^2 \right)}} |\alpha\rangle. \tag{7.2.7}$$

Once obtained the outcome α measuring the ancillary qubit, we can perform a measurement on the index register, the probability to obtain the outcome i is:

$$\mathbb{P}(i|\alpha) = \langle \Psi_\alpha | Q_i \Psi_\alpha \rangle \quad \text{with} \quad Q_i = |i\rangle\langle i| \otimes \mathbb{I}_{n,a}, \tag{7.2.8}$$

where $\mathbb{I}_{n,a}$ is the identity operator on the Hilbert space $\mathsf{H}_n^{\otimes 2} \otimes \mathsf{H}_a$ and $|i\rangle\langle i|$ is the projector onto the subspace of H_{index} spanned by $|i\rangle$. Therefore:

$$\mathbb{P}(i|\alpha) = \frac{(\langle \mathbf{x}_i | \langle \mathbf{x} | + (-1)^\alpha \langle \mathbf{x} | \langle \mathbf{x}_i |)(|\mathbf{x}_i\rangle |\mathbf{x}\rangle + (-1)^\alpha |\mathbf{x}\rangle |\mathbf{x}_i\rangle)}{2(N + (-1)^\alpha \sum_{i=0}^{N-1} |\langle \mathbf{x} | \mathbf{x}_i \rangle|^2)} \tag{7.2.9}$$

$$= \frac{2 + 2(-1)^\alpha \langle \mathbf{x}_i | \mathbf{x} \rangle \langle \mathbf{x} | \mathbf{x}_i \rangle}{2(N + (-1)^\alpha \sum_{i=0}^{N-1} |\langle \mathbf{x} | \mathbf{x}_i \rangle|^2)}$$

$$= \frac{1 + (-1)^\alpha |\langle \mathbf{x} | \mathbf{x}_i \rangle|^2}{N + (-1)^\alpha \sum_{i=0}^{N-1} |\langle \mathbf{x} | \mathbf{x}_i \rangle|^2}.$$

Now define the following quantity:

$$\mathbb{Q}(i) := \mathbb{P}(i|0) - \mathbb{P}(i|1) = \frac{2(|\langle \mathbf{x} | \mathbf{x}_i \rangle|^2 - C)}{N(1 - C^2)}, \tag{7.2.10}$$

where $C = \frac{1}{N} \sum_i |\langle \mathbf{x} | \mathbf{x}_i \rangle|^2$ is a constant, so $\mathbb{Q}$ is proportional to the square cosine similarity between $\mathbf{x}_i$ and $\mathbf{x}$ that is the *fidelity* between the quantum states $|\mathbf{x}\rangle$ and $|\mathbf{x}_i\rangle$. Since the vectors have positive entries, sampling the index register one selects the indexes with the highest value of $\mathbb{Q}$ corresponding to the closest vectors to $\mathbf{x}$.

Classical execution of a kNN algorithm requires the computation of the distance (or the similarity) between the test instance and all the elements of the training

Input: training set $X = \{\mathbf{x}_i, y_i\}_{i=0,...,N-1}$, unclassified instance $\mathbf{x}$.
Result: label y of $\mathbf{x}$.

1 **repeat**
2   initialize the register $\mathsf{H}_{index} \otimes \mathsf{H}_n \otimes \mathsf{H}_n$ and an ancillary qubit a in the state (7.2.2);
3   perform the SWAP test on the n-qubit registers controlled by the qubit a;
4   measure the qubit a;
5   measure the index register;
6 **until** *desired accuracy on the estimation of* $\mathbb{Q}$;
7 select the K indexes $\{i_1, ..., i_K\}$ with highest $\mathbb{Q}$;
8 determine y by majority voting over $X \leftarrow \{\mathbf{x}_{i_k}, y_{i_k}\}_{k=1,...,K}$;
9 **return** y;

Algorithm 8: *Quantum kNN.*

set to find the k nearest neighbors. In this quantum kNN, only the closest training points present a high probability to be detected by the measurements, in other words the far training points consume fewer resources. In the classical case, the classification of a new input requires time $O(Nd)$, where d is the dimension of the feature space and N is the number of training vectors. The quantum kNN (Algorithm 8) requires $O(\log(Nd))$ steps to retrieve the initial state (7.2.2) from the QRAM and $\log d$ Fredkin gates to implement the SWAP test, then the overall time complexity is $O(\log(Nd^2))$. The number of qubits required by the four registers is $2 \log d + \log N + 1$ then the space complexity of the quantum kNN is $O(\log(Nd))$.

7.3 Quantum support vector machine

The *support vector machine* (SVM) is a supervised learning algorithm for binary classification. In particular SVM provides a *decision boundary* (a hyperplane in the feature space) separating the considered classes such that its distance to each class is as large as possible [CV95, Ce20].

Given a training set $\{(\mathbf{x}_i, y_i)\}_{i=1,...,N}$ where $\mathbf{x}_i \in \mathbb{R}^d$ and $y_i \in \{\pm 1\}$, the aim of the SVM is to find a hyperplane with equation $\mathbf{w} \cdot \mathbf{x} + b = 0$, with $\mathbf{w} \in \mathbb{R}^d$ and $b \in \mathbb{R}$, that correctly classifies the data maximizing the distance between the hyperplane and the closest data points. These points are called *support vectors* and lie on two parallel hyperplanes with equations $\mathbf{w} \cdot \mathbf{x} + b = \pm 1$. The problem can be formulated in the following terms:

$$\operatorname*{argmin}_{\mathbf{w},b} \frac{\| \mathbf{w} \|^2}{2} \quad \text{s.t.} \quad y_i(\mathbf{w} \cdot \mathbf{x}_i + b) \geq 1 \quad \forall i = 1, ..., N, \qquad (7.3.1)$$

that is task of standard quadratic programming. The problem can be reformulated by means the *Karush-Kuhn-Tucker multipliers* $\alpha = \{\alpha_1, ..., \alpha_N\}$ defining the

Lagrangian:

$$\mathcal{L}(\mathbf{w}, b, \alpha) = \frac{1}{2} \parallel \mathbf{w} \parallel^2 - \sum_{i=1}^{N} \alpha_i [y_i (\mathbf{w} \cdot \mathbf{x}_i + b) - 1], \qquad (7.3.2)$$

that must be maximized with respect to α with the following constraints:

$$\frac{\partial \mathcal{L}}{\partial \mathbf{w}} = \mathbf{w} - \sum_{i=1}^{N} \alpha_i y_i \mathbf{x}_i = 0, \qquad \frac{\partial \mathcal{L}}{\partial b} = \sum_{i=1}^{N} \alpha_i y_i = 0, \qquad \alpha_i \geq 0 \quad \forall i = 1, ..., N. \quad (7.3.3)$$

Equation (7.3.2) can be written as:

$$\mathcal{L}(\mathbf{w}, b, \alpha) = \frac{1}{2} \mathbf{w} \cdot \mathbf{w} - \mathbf{w} \cdot \sum_{i=1}^{N} \alpha_i y_i \mathbf{x}_i - b \sum_{i=1}^{N} \alpha_i y_i + \sum_{i=1}^{N} \alpha_i, \qquad (7.3.4)$$

then by a substitution, we obtain the Lagrangian:

$$\mathcal{L}_d(\alpha) = \sum_{i=1}^{N} \alpha_i - \frac{1}{2} \sum_{i,j=1}^{N} \alpha_i \alpha_j y_i y_j \, \mathbf{x}_i \cdot \mathbf{x}_j \qquad (7.3.5)$$

for the formulation of the dual problem that does not depend on $\mathbf{w}$ and b:

$$\arg \max_{\alpha} \mathcal{L}_d(\alpha) \quad \text{s.t.} \quad \sum_{i=1}^{N} \alpha_i y_i = 0 \quad \text{and} \quad \alpha \geq 0 \;\; \forall i = 1, ...N. \qquad (7.3.6)$$

Once maximized $\mathcal{L}_d$, the weights $\mathbf{w}$ and b can be recovered by $\mathbf{w} = \sum_{i=1}^{N} \alpha_i y_i \mathbf{x}_i$ and $b = y_i - \mathbf{w} \cdot \mathbf{x}_i$ for any i such that $\alpha_i > 0$ (the corresponding $\mathbf{x}_i$ are the support vectors). From (7.4.3), we can define the *kernel function* as $\kappa(\mathbf{x}_i, \mathbf{x}_j) := \mathbf{x}_i \cdot \mathbf{x}_j$. There are more complicated kernel functions to generalize the SVM to non-linear classifications that we introduce below.

Solving the dual problem (7.3.6) requires the calculation of $\frac{N(N-1)}{2}$ dot products (a dot product takes time $O(d)$) and the quadratic programming itself takes time $O(N^3)$. Thus the classical SVM takes time $O(N^2(N + d))$ to provide a binary classifier that assigns the label y to $\mathbf{x}$ according to:

$$y(\mathbf{x}) = \operatorname{sgn} \left(\sum_{i=1}^{N} \alpha_i \kappa(\mathbf{x}_i, \mathbf{x}) + b \right). \qquad (7.3.7)$$

A quantum implementation of SVM can be defined within the amplitude encoding of data. The retrieval from a QRAM requires $O(N \log d)$ operations. The computation of the kernel can be performed by the estimations of $|\langle \mathbf{x}_i | \mathbf{x}_j \rangle|^2$ with the SWAP test. The running time of this quantum SVM is $O(N^2(N + \log d))$.

Here we summarize the quantum version of SVM proposed in [Re14] based on efficient preparation and inversion of the *kernel matrix* that is the $N \times N$ matrix K given by $k_{ij} = \kappa(\mathbf{x}_i, \mathbf{x}_j)$, for the quantum formulation of the *least-squares SVM*. In fact the quadratic programming problem can be reformulated in terms of a linear programming introducing *slack variables* ξ_j replacing the inequality constraints in (7.3.1) with equality constraints:

$$y_i(\mathbf{w} \cdot \mathbf{x}_i + b) = 1 - \xi_i \qquad i = 1, ..., N, \qquad (7.3.8)$$

ξ_i quantifies the deviation of a data instance from the margin. In addition to the constraints (7.3.8), we have to minimize a penalty term, so the considered optimization problem is:

$$\arg\min_{\mathbf{w}, b, e} \frac{1}{2} \parallel \mathbf{w} \parallel^2 + \frac{\gamma}{2} \sum_{i=1}^{N} \xi_i^2 \quad s.t. \quad y_i(\mathbf{w}_i \cdot \mathbf{x}_i + b) = 1 - \xi_i \quad \forall i = 1, ..., N. \ (7.3.9)$$

The tuning parameter $\gamma > 0$ represents the weight of misclassification in deciding the hyperplane. The associated Lagrangian is:

$$\mathfrak{L}(\mathbf{w}, b, \xi, \alpha) = \frac{1}{2} \parallel \mathbf{w} \parallel^2 + \frac{\gamma}{2} \sum_{i=1}^{N} \xi_i - \sum_{i=1}^{N} \alpha_i(\mathbf{w} \cdot \mathbf{x}_i + b + y_i \xi_i - y_i). \quad (7.3.10)$$

and the optimality conditions are:

$$
\begin{cases}
\frac{\partial \mathfrak{L}}{\partial \mathbf{w}} = \mathbf{w} - \sum_{i=1}^{N} \alpha_i \mathbf{x}_i = 0 \\[2mm]
\frac{\partial \mathfrak{L}}{\partial b} = -\sum_{i=1}^{N} \alpha_i = 0 \\[2mm]
\frac{\partial \mathfrak{L}}{\partial \xi_i} = \gamma \xi_i - \alpha_i y_i = 0 \\[2mm]
\frac{\partial \mathfrak{L}}{\partial \alpha_i} = \mathbf{w} \cdot \mathbf{x}_i + b + y_i \xi_i - y_i = 0 \, .
\end{cases}
\qquad (7.3.11)
$$

Therefore the optimization problem reduces to the resolution of the linear system:

$$
\begin{cases}
\sum_{i=1}^{N} \alpha_i = 0 \\[2mm]
\sum_{i=1}^{N} \alpha_i \mathbf{x}_i \cdot \mathbf{x}_j + b + \gamma^{-1} \alpha_j = y_j \qquad j = 1, ..., N.
\end{cases}
\qquad (7.3.12)
$$

From the solution (b, α) of (7.3.12), one can deduce $\mathbf{w}$ and ξ_i by $\mathbf{w} = \sum_{i=1}^{N} \alpha_i \mathbf{x}_i$ and $\xi_i = \gamma^{-1} \alpha_i$ respectively. The linear system in the matrix form is:

$$F \begin{pmatrix} b \\ \alpha \end{pmatrix} \equiv \begin{pmatrix} 0 & 1^T \\ 1 & K + \gamma^{-1}\mathbb{I} \end{pmatrix} \begin{pmatrix} b \\ \alpha \end{pmatrix} = \begin{pmatrix} 0 \\ y \end{pmatrix}, \qquad (7.3.13)$$

where K is the kernel matrix, $y = (y_1, ..., y_N)^T$, $1 = (1, ..., 1)^T$ and $\mathbb{I}$ is the $N \times N$ identity matrix. Once calculated the kernel matrix, the classical complexity of the least-squares SVM is $O(N^3)$ as for the quadratic programming.

The quantum version of the LS-SVM proposed in [Re14] is based on the preparation of the normalized kernel matrix $\hat{K} = K/\mathrm{tr}K$ as a density matrix and on the application of an efficient quantum algorithm to solve (7.3.13), the *HHL algorithm* [Ha09]. The procedure starts with the preparation of the quantum state:

$$|\Psi\rangle = \frac{1}{\sqrt{C}} \sum_{i=1}^{N} \| \mathbf{x}_i \| |i\rangle|\mathbf{x}_i\rangle \quad \text{with} \quad C = \sum_{i=1}^{N} \| \mathbf{x}_i \|^2, \qquad (7.3.14)$$

that can be retrieved from the QRAM in $O(\log[Nd])$ operations. Discarding the data set register, the reduced state in the first register is given by the partial trace:

$$\mathrm{tr}_2(|\Psi\rangle\langle\Psi|) = \frac{1}{C} \sum_{k=1}^{d} \sum_{i,j=1}^{N} \| \mathbf{x}_i \|\| \mathbf{x}_j \| \langle k|\mathbf{x}_i\rangle\langle\mathbf{x}_j|k\rangle |i\rangle\langle j|$$

$$= \frac{1}{C} \sum_{i,j=1}^{N} \| \mathbf{x}_i \|\| \mathbf{x}_j \| \langle\mathbf{x}_j|\mathbf{x}_i\rangle|i\rangle\langle j| = \frac{K}{\mathrm{tr}K}. \qquad (7.3.15)$$

The preparation of the kernel matrix resembles the preparation of the normalized covariance matrix as a density matrix discussed in Section 6.1. In the same section we have also described the efficient exponentiation of a density matrix, we need to exponentiate the normalized matrix $\hat{F} = F/\mathrm{tr}F$ and apply the phase estimation algorithm. Let $|y\rangle$ be the quantum state encoding $(0, y)^T$ into the amplitudes with respect to the basis $\{|i\rangle\}_{i=0,1,...,N}$. Consider the main register storing $|y\rangle$, an additional n-qubit register to store the estimation of the eigenvalues of $\hat{F}$ and an ancillary qubit prepared in the initial state $|\Psi_0\rangle = |y\rangle|0\rangle^{\otimes n}|0\rangle$, the phase estimation algorithm returns the state:

$$|\Psi_1\rangle = \sum_{i=0}^{N} \langle\psi_i|y\rangle|\psi_i\rangle\left|\hat{\lambda}_i\right\rangle|0\rangle, \qquad (7.3.16)$$

where $\{|\psi_i\rangle\}_{i=0,\ldots,N}$ are the eigenvectors of $\hat{F}$ and $\hat{\lambda}_i$ are the estimations of its eigenvalues. Then a controlled rotation is applied on the ancillary qubit controlled by the n-qubit register producing the state:

$$|\Psi_2\rangle = \sum_{i=0}^{N} \langle\psi_i|y\rangle |\psi_i\rangle \big|\hat{\lambda}_i\big\rangle \left(\sqrt{1 - \frac{1}{\hat{\lambda}_i^2}} |0\rangle + \frac{1}{\hat{\lambda}_i} |1\rangle \right). \qquad (7.3.17)$$

After a measurement process on the ancillary qubit with outcome 1, the state in the main register is:

$$|\Psi_f\rangle = \sum_{i=0}^{N} \frac{\langle\psi_i|y\rangle}{\hat{\lambda}_i} |\psi_i\rangle \simeq \sum_{i=0}^{N} \frac{\langle\psi_i|y\rangle}{\lambda_i} |\psi_i\rangle. \qquad (7.3.18)$$

Since:

$$\hat{F}|\Psi_f\rangle \simeq \sum_{i=0}^{N} \lambda_i \frac{\langle\psi_i|y\rangle}{\lambda_i} |\psi_i\rangle = \sum_{i=0}^{N} \langle\psi_i|y\rangle |\psi_i\rangle = |y\rangle, \qquad (7.3.19)$$

the amplitudes of $|\Psi_f\rangle$ with respect to the basis $\{|i\rangle\}_{i=0,\ldots,N}$ are an estimation of the parameters of the SVM, that is, the final state is close (in the sense of phase estimation algorithm) to:

$$|b, \alpha\rangle = \frac{1}{\sqrt{D}} \left(b|0\rangle + \sum_{i=1}^{N} \alpha_i |i\rangle \right), \qquad (7.3.20)$$

where $D = b^2 + \sum_{i=1}^{N} \alpha_i^2$. The estimation of the state $|b, \alpha\rangle$ represents the *training* of the quantum SVM. Once trained the quantum SVM, the classification a new input encoded in $|\mathbf{x}\rangle$ is done as follows: From $|b, \alpha\rangle$, let us construct the state encoding the training set:

$$|\phi\rangle = \frac{1}{\sqrt{C_\phi}} \left(b|0\rangle|0\rangle + \sum_{i=1}^{N} \alpha_i \parallel \mathbf{x}_i \parallel |i\rangle|\mathbf{x}_i\rangle \right) \qquad (7.3.21)$$

where $C_\phi = b^2 + \sum_i \alpha_i^2 \parallel \mathbf{x}_i \parallel^2$. Then let us construct the query state:

$$|\varphi_\mathbf{x}\rangle = \frac{1}{\sqrt{C_\mathbf{x}}} \left(|0\rangle|0\rangle + \sum_{i=1}^{N} \parallel \mathbf{x} \parallel |i\rangle|\mathbf{x}\rangle \right), \qquad (7.3.22)$$

with $C_\mathbf{x} = 1 + N \parallel \mathbf{x} \parallel^2$. Using an ancillary qubit, we can construct the state $\frac{1}{\sqrt{2}}(|0\rangle|\phi\rangle + |1\rangle|\varphi_\mathbf{x}\rangle)$ and perform the SWAP test between the ancillary part and the qubit state $|-\rangle = \frac{1}{\sqrt{2}}(|0\rangle - |1\rangle)$ estimating the inner product $\langle\Phi|\Phi\rangle$ where Φ is the non-normalized vector:

$$\Phi := \frac{1}{\sqrt{2}}((\langle-|0\rangle|\phi\rangle + \langle-|1\rangle|\varphi_{\mathbf{x}}\rangle) = \frac{1}{2}(|\phi\rangle - |\varphi_{\mathbf{x}}\rangle). \qquad (7.3.23)$$

Since $\langle\Phi|\Phi\rangle = \frac{1}{2}(1 - \langle\phi|\varphi_{\mathbf{x}}\rangle)$ and

$$\langle\phi|\varphi_{\mathbf{x}}\rangle = \frac{1}{\sqrt{C_\phi C_{\mathbf{x}}}} \left(b + \sum_{i=1}^{N} \alpha_i \parallel \mathbf{x}_i \parallel \parallel \mathbf{x} \parallel \langle\mathbf{x}_i|\mathbf{x}\rangle \right) , \qquad (7.3.24)$$

the SWAP test allows to perform the binary classification. More precisely if the probability of success (that is the probability to measure 0 on the qubit that controls the Fredkin gate in the SWAP test) is $\mathbb{P} = \langle\Phi|\Phi\rangle < 1/2$ then $\mathbf{x}$ is classified as $+1$, otherwise as -1.

The time required by the training of the quantum SVM is given by the time access to retrieve the kernel matrix and the simulation of $e^{-i\hat{K}\Delta t}$ ($\hat{K}$ is the non-sparse contribution to $\hat{F}$) then the time complexity is $O(\log[Nd])$ while the classical training time is $O(N^2(N+d))$. The classification with the classical SVM takes time $O(Nd)$, in the quantum case the classification takes a constant time and requires $O(\epsilon^{-2})$ repetitions for an accuracy of ϵ. In [Re14] there is the complete discussion about the error estimation related to the approximations due to the phase estimation and the density matrix exponentiation.

If the data set cannot be linearly separated, we can define a non-linear mapping into a higher dimensional space $\phi : \mathbb{R}^d \to \mathbb{R}^{\tilde{d}}$, with $\tilde{d} > d$, such that the transformed data $\phi(\mathbf{x}_i)$ can be separated by a hyperplane in $\mathbb{R}^{\tilde{d}}$. Thus the considered kernel function is:

$$\kappa(\mathbf{x}_i, \mathbf{x}_j) = \phi(\mathbf{x}_i) \cdot \phi(\mathbf{x}_j). \qquad (7.3.25)$$

In the case of a polynomial kernel like $\kappa(\mathbf{x}_i, \mathbf{x}_j) = (\mathbf{x}_i \cdot \mathbf{x}_j)^a$, the quantum encoding offers an efficient way to implement it: It is sufficient taking the a-times tensor product of the states encoding the data: $|\phi(\mathbf{x}_i)\rangle = |\mathbf{x}_i\rangle^{\otimes a}$. Therefore we simply have $\langle\phi(\mathbf{x}_i)|\phi(\mathbf{x}_j)\rangle = \langle\mathbf{x}_i|\mathbf{x}_j\rangle^a$ and the discussed quantum SVM machine can be implemented in the higher dimensional space.

7.4   SVM training with a quantum annealer

A parametric predictive model can be trained by a quantum annealer if the corresponding optimization problem can be formulated into the QUBO form and represented into the annealer architecture. In general, the task of optimizing real parameters $\alpha_1, ..., \alpha_N$ must be formulated as a binary optimization problem of the form:

$$\underset{\mathbf{z} \in \{0,1\}^n}{\arg\min} \ \mathbf{z}^T Q \mathbf{z}, \tag{7.4.1}$$

where Q is a real upper-triangular matrix. The real variables $\boldsymbol{\alpha}$ can be encoded into binary variables $\mathbf{z}$ as follows:

$$\mathbb{R} \ni \alpha_i = \sum_{k=0}^{M-1} B^k \mathbf{z}_{Mi+k}, \tag{7.4.2}$$

where z_{Mi+k} are the binary variables, M is the number of binary variables used to encode α_i and B is the base used for the encoding. As in the case of the SVM discussed below, the optimization problem for the model training can be constrained so a penalty term may be introduced into the QUBO formulation.

Since the relevance of support vector machine as a kernel method, let us summarize the training procedure of a SVM by means of quantum annealing [Wi20]. The quadratic programming problem (7.3.6), formulated in the previous section, can be directly re-written as the constrained minimization problem:

$$\underset{\boldsymbol{\alpha}}{\arg\min} \left[\frac{1}{2} \sum_{i,j=1}^{N} \alpha_i \alpha_j y_i y_j \, k(\mathbf{x}_i, \mathbf{x}_j) - \sum_{i=1}^{N} \alpha_i \right] \ \text{s.t.} \ \sum_{i=1}^{N} \alpha_i y_i = 0 \ \text{and} \ \alpha \ge 0 \ \forall i = 1, ... N \tag{7.4.3}$$

where we consider an arbitrary kernel function k. The constrained problem (7.4.3) can be formulated as the unconstrained minimization of the following quadratic form:

$$E(\mathbf{z}) = \sum_{i,j=0}^{N-1} \sum_{k,q=0}^{M-1} z_{Mi+k} \, \widetilde{Q}_{Mi+k,Mj+q} \, z_{Mj+q}, \tag{7.4.4}$$

where the real symmetric $NM \times NM$ matrix $\widetilde{Q}$ is defined by:

$$\widetilde{Q}_{Mi+k,Mj+q} = \frac{1}{2} B^{k+q} y_i y_j \left[k(\mathbf{x}_i, \mathbf{x}_j) + \xi \right] - \delta_{ij} \delta_{kq} B^k. \tag{7.4.5}$$

The constrain is included by means of a squared penalty term with the multiplier ξ. The upper-triangular QUBO matrix is given by:

$$Q_{ij} := \widetilde{Q}_{ij} + \widetilde{Q}_{ji} \ \ i < j, \tag{7.4.6}$$

$$Q_{ii} := \widetilde{Q}_{ii}.$$

The case of a Gaussian kernel is particularly suitable for the embedding into an annealer sparse architecture because the property $k(\mathbf{x}_i, \mathbf{x}_j) \simeq 0$ as $\| \mathbf{x}_i - \mathbf{x}_j \| \gg 1$ allows to reduce the number of qubit couplings. Te results reported in [Wi20] show that the quantum annealer produces not just the global optimum for the training data, but a distribution of many reasonably good, close-to-optimal solutions to the given optimization problem. Then the samplings performed by quantum annealing provide an ensemble of classifiers rather than a single trained SVM, a combination of these classifiers has the potential to generalize better to the test data.

7.5   Quantum-inspired classification

The so-called *quantum-inspired machine learning* is based on particular kinds of data representation and processing defined by means of mathematical objects from the quantum formalism that do not necessary represent physical quantum systems. Within this paradigm, the mathematical structures of the quantum theory are used to devise novel machine learning algorithms intended for classical hardware.

In [Se19], there is the proposal of a quantum-inspired classification algorithm based on a generalization of the Helstrom state discrimination [He69]. Let us focus on the case of binary classification of n-dimensional complex feature vectors, the algorithm is based on the following three ingredients:

1) quantum encoding of the feature vectors into density operators;
2) construction of the quantum centroids of the two classes C_1 and C_2 of training points:

$$\rho_i := \frac{1}{|C_i|} \sum_{\mathbf{x} \in C_i} \rho_{\mathbf{x}} \qquad i = 1, 2, \qquad (7.5.1)$$

where $\rho_{\mathbf{x}}$ is the quantum state encoding the training vector $\mathbf{x}$;
3) application of the Helstrom discrimination on the two quantum centroids in order to assign a label to a new data instance.

Let us briefly describe the notion of quantum state discrimination. Given a set of arbitrary quantum states with respective a priori probabilities $X = \{(\rho_1, p_1), ..., (\rho_N, p_N)\}$, assume to have a source that emits quantum systems whose states obey to the statistic described by X, in general there is no a measurement process over the emitted quantum systems that discriminates the states without errors. In some particular cases the states can be exactly discriminated, for example if we have a set of orthogonal pure states $\{|\psi_1\rangle, ..., |\psi_N\rangle\}$ we can discriminate them without errors by means of the corresponding measurement process $\{|\psi_i\rangle\langle\psi_i|\}_{i=1,...,N}$. In the case of a general set X, the probability of a successful state discrimination performing the measurement E described by the PVM $\{P_i\}_i$ is:

$$\mathbb{P}_E(X) = \sum_{i=1}^{N} p_i \text{tr}(P_i \rho_i). \tag{7.5.2}$$

An interesting and useful task is finding the optimal measurement that maximizes the probability (7.5.2). There is a complete characterization of the optimal measurement E_{opt} for $X = \{(\rho_1, p_1), (\rho_2, p_2)\}$. E_{opt} can be constructed as follows [He69]: Let $\Lambda := p_1 \rho_1 - p_2 \rho_2$ be the *Helstrom observable* whose positive and negative eigenvalues are respectively collected in the sets D_+ and D_-. Consider the two orthogonal projectors:

$$P_\pm := \sum_{\lambda \in D_\pm} P_\lambda, \tag{7.5.3}$$

where P_λ projects onto the eigenspace of λ. The measurement $E_{opt} := \{P_+, P_-\}$ maximizes the probability (7.5.2) that attains the *Helstrom bound* $h_b(\rho_1, \rho_2) = p_1 \text{tr}(P_+ \rho_1) + p_2 \text{tr}(P_- \rho_2)$.

Helstrom quantum state discrimination can be used to implement a binary classifier [Se19]. Let $\{(\mathbf{x}_1, y_1), ..., (\mathbf{x}_M, y_M)\}$ be a training set with $y_i \in \{1, 2\}$ $\forall i = 1, ..., M$. Once selected a quantum encoding, one can construct the quantum centroids ρ_1 and ρ_2 as in (7.5.1) of the two classes $C_{1,2} = \{\mathbf{x}_i : y_i = 1, 2\}$. Let $\{P_+, P_-\}$ be the Helstrom measurement defined by the set $X = \{(\rho_1, p_1), (\rho_2, p_2)\}$ where the probabilities attached to the centroids are $p_{1,2} = \frac{|C_{1,2}|}{|C_1|+|C_2|}$. The *Helstrom classifier* applies the optimal measurement for the discrimination of the two quantum centroids to assign the label y to a new data instance $\mathbf{x}$, encoded into the state $\rho_\mathbf{x}$, as follows:

$$y(\mathbf{x}) = \begin{cases} 1 & \text{if} \quad \text{tr}(P_+ \rho_\mathbf{x}) \geq \text{tr}(P_- \rho_\mathbf{x}) \\ 2 & \text{otherwise} \end{cases}. \tag{7.5.4}$$

A strategy to increase the accuracy in classification is given by the construction of the tensor product of k copies of the quantum centroids $\rho_{1,2}^{\otimes k}$ enlarging the Hilbert space where data are encoded. The corresponding Helstrom measurement is $\{P_+^{\otimes k}, P_-^{\otimes k}\}$ and the Helstrom bound satisfies [Giu21]:

$$h_b(\rho_1^{\otimes k}, \rho_2^{\otimes k}) \leq h_b\left(\rho_1^{\otimes(k+1)}, \rho_2^{\otimes(k+1)}\right) \qquad \forall k \in \mathbb{N}. \tag{7.5.5}$$

Enlarging the Hilbert space of the quantum encoding, one increases the Helstrom bound obtaining a more accurate classifier. The computational cost of such a kernel trick is clearly high, however in the case of real input vectors the space can be enlarged saving time and space by means of a quantum encoding into *Bloch vectors* [LP21, LP22].

Chapter 8

Quantum pattern recognition

In this chapter we introduce the quantum implementation of an associative memory based on a modification of the Grover algorithm. Then we review the application of the quantum Fourier transform to pattern recognition and an adiabatic algorithm to retrieve binary patterns from a quantum memory.

8.1 Quantum associative memory

A random access memory is address-oriented, that is, the retrieval of information from the storage is done by means of the address to a memory location. Thus, incomplete or noisy addresses cannot be used to retrieve information. An associative memory is a structure that permits to retrieve information using a *pattern* instead of an address, so the retrieval can be done on the basis of partial knowledge of the memory content and without knowing a storage location.

In the classical case, an associative memory can be realized by means of collective computation on neural networks, the most prominent example is the *Hopfield model* [Ho82]. However, classical models suffer from a severe capacity storage, for instance the maximum number of binary patterns that can be stored in a Hopfield network of n neurons is $N_{max} \simeq 0.14n$ and more generally $N_{max} = O(n)$ [Mü90].

In a quantum memory made by n qubits, we can store all the binary patterns of length n. In order to use the significant storing capacity of a quantum memory to realize an associative memory, we need an efficient procedure of *pattern recognition*. An associative memory can be obtained by a modification of Grover's algorithm. Let us reconsider the Grover's search in Algorithm 9 where $U_f|x\rangle := (-1)^{f(x)}|x\rangle$ is the oracle operator marking the solution and $U_G := H^{\otimes n}(2|0\rangle\langle 0| - \mathbb{I})H^{\otimes n}$ is the Grover's diffusion operator. The composition $G := U_G U_f$ is the Grover's iteration. The application of Grover to retrieve a binary pattern from a memory cannot be

© The Author(s), under exclusive license to Springer Nature Singapore Pte Ltd. 2023
D. Pastorello, *Concise Guide to Quantum Machine Learning*,
Machine Learning: Foundations, Methodologies, and Applications,
https://doi.org/10.1007/978-981-19-6897-6_8

directly realized, in fact Grover's algorithm searches in the set of all the binary strings of length n. We need a modification in order to apply Grover for the retrieval from a collection of binary patterns of length n that does not include every binary pattern of length n.

> **Input:** Unsorted database X of N items, oracle function $f : X \to \{0, 1\}$ marking $\hat{x}$.
> **Result:** target item $\hat{x}$
> 1 $\Psi \leftarrow H^{\otimes n}|0\rangle$;
> 2 **Repeat** $\frac{\pi}{4}\sqrt{N}$ times:
> 3 $|\Psi\rangle \leftarrow U_f|\Psi\rangle$;
> 4 $|\Psi\rangle \leftarrow U_G|\Psi\rangle$;
> 5 measure

Algorithm 9: *Grover's algorithm.*

Consider a collection of binary patterns of length n that we need to store and from which we need to retrieve:

$$\mathcal{P} = \{p^{(1)}, ..., p^{(N)}\} \qquad N < 2^n.$$

Let us assume to encode the patterns into the basis states of a n-qubit register: $p^{(k)} \mapsto |p^{(k)}\rangle$. The first idea for a quantum associative memory is storing the binary patterns of $\mathcal{P}$ into the superposition:

$$|M\rangle = \frac{1}{\sqrt{N}} \sum_{i=1}^{N} |p^{(k)}\rangle$$

and run Grover on $|M\rangle$ to retrieve a target binary pattern. There are two fundamental issues to manage: Implementation of an efficient preparation of $|M\rangle$ and checking whether Grover works well starting from a state that is not the superposition of all the basis states.

For the storing of the N patterns, three qubit registers are required [VM00]: An input n-qubit register to upload the patterns, a utility two-qubit register and a memory n-qubit register to store the uploaded patterns. The patterns are uploaded one by one, the initial state for the uploading of the first pattern $p^{(1)} \equiv p = p_1 \dots p_n$ is:

$$|\Psi_0\rangle = |p_1 \dots p_n\rangle_{in}|01\rangle_u|0 \dots 0\rangle_m. \tag{8.1.1}$$

The storing procedure of p is given by the following two steps:

1. $$|\Psi_1\rangle = \prod_{j=1}^{n} \mathsf{CNOT}_{in_j m_j}|\Psi_0\rangle = |p_1 \cdots p_n\rangle_{in}|01\rangle_u|p_1 \cdots p_n\rangle_m$$

2. $\quad |\Psi_2\rangle = S_{u_2}^{(N)}|\Psi_1\rangle = |p\rangle_{in}|0\rangle_{u_1}\left(\frac{1}{\sqrt{N}}|0\rangle_{u_2} + \sqrt{\frac{N-1}{N}}|1\rangle_{u_2}\right)|p\rangle_m$

$$= \frac{1}{\sqrt{N}}|p\rangle_{in}|00\rangle_u|p\rangle_m + \sqrt{\frac{N-1}{N}}|p\rangle_{in}|01\rangle_u|p\rangle_m$$

where $\mathsf{CNOT}_{in_j m_j}$ denotes the action of a CNOT gate on the jth qubit of the memory register controlled by the jth qubit of the input register. This operation copies the pattern into the memory register. In the step 2, we have used the action of the gate:

$$S^{(i)} = \begin{pmatrix} \sqrt{\frac{i-1}{i}} & \frac{1}{\sqrt{i}} \\ -\frac{1}{\sqrt{i}} & \sqrt{\frac{i-1}{i}} \end{pmatrix} \qquad i = 1, ..., N \qquad (8.1.2)$$

on the second qubit of the utility register (denoted by the subscript), for $i = N$. The action of $S_{u_2}^{(N)}$ create a superposition of two terms: the first one is called the *storing component* and it is used to encode stored patterns, the second one is called *processing component* and it is used to process a new pattern. In order to upload a new pattern we have to reset the memory register in the processing component of $|\Psi_2\rangle$:

$$\prod_{j=n}^{1} \mathsf{CNOT}_{in_j u_2 m_j}|\Psi_2\rangle = \frac{1}{\sqrt{N}}|p\rangle_{in}|00\rangle_u|p\rangle_m + \sqrt{\frac{N-1}{N}}|p\rangle_{in}|01\rangle_u|0\cdots 0\rangle_m, \quad (8.1.3)$$

where $\mathsf{CNOT}_{in_j u_2 m_j}$ denotes a CNOT gate on the j-th qubit of the memory register controlled by the jth qubit of the input register and the second utility qubit. Now the system is ready to receive the second pattern $p^{(2)} \equiv q$ in the input register:

$$|\Phi_0\rangle = \frac{1}{\sqrt{N}}|q\rangle_{in}|00\rangle_u|p\rangle_m + \sqrt{\frac{N-1}{N}}|q\rangle_{in}|01\rangle_u|0\cdots 0\rangle_m$$

The storing procedure of q is given by the following steps:

1. $|\Phi_1\rangle = \prod_{j=1}^{n}\mathsf{CNOT}_{in_j u_2 m_j}|\Phi_0\rangle = \frac{1}{\sqrt{N}}|q\rangle_{in}|00\rangle_u|p\rangle_m + \sqrt{\frac{N-1}{N}}|q\rangle_{in}|01\rangle_u|q\rangle_m$

2. $|\Phi_2\rangle = \prod_{j=1}^{n}\mathsf{NOT}_{m_j}\mathsf{CNOT}_{in_j m_j}|\Phi_1\rangle = \frac{1}{\sqrt{N}}|q\rangle_{in}|00\rangle_u|\overline{p\oplus q}\rangle_m + \sqrt{\frac{N-1}{N}}|q\rangle_{in}|01\rangle_u|1\rangle_m$

3. $|\Phi_3\rangle = \mathsf{CNOT}_{m_1\cdots m_n u_1}|\Phi_2\rangle = \frac{1}{\sqrt{N}}|q\rangle_{in}|00\rangle_u|\overline{p\oplus q}\rangle_m + \sqrt{\frac{N-1}{N}}|q\rangle_{in}|11\rangle_u|1\rangle_m$

4. $|\Phi_4\rangle = \mathsf{CS}_{u_1 u_2}^{(N-1)}|\Phi_3\rangle = \frac{1}{\sqrt{N}}|q\rangle_{in}|00\rangle_u|\overline{p\oplus q}\rangle_m + \sqrt{\frac{N-1}{N}}|q\rangle_{in}|1\rangle_{u_1}\left(\frac{1}{\sqrt{N-1}}|0\rangle_{u_2} + \sqrt{\frac{N-2}{N-1}}|1\rangle_{u_2}\right)|1\rangle_m$

5. $|\Phi_5\rangle = \text{CNOT}_{m_1\cdots m_n u_1}|\Phi_4\rangle = \frac{1}{\sqrt{N}}|q\rangle_{in}|00\rangle_u|\overline{p\oplus q}\rangle_m + \sqrt{\frac{N-1}{N}}|q\rangle_{in}|0\rangle_{u_1}\left(\frac{1}{\sqrt{N-1}}|0\rangle_{u_2} + \sqrt{\frac{N-2}{N-1}}|1\rangle_{u_2}\right)|1\rangle_m$

6. $|\Phi_6\rangle = \prod_{j=n}^{1}\text{CNOT}_{in_j m_j}\text{NOT}_{m_j}|\Phi_5\rangle = \frac{1}{\sqrt{N}}|q\rangle_{in}|00\rangle_u|p\rangle_m + \frac{1}{\sqrt{N}}|q\rangle_{in}|00\rangle_u|q\rangle_m +$
$\sqrt{\frac{N-2}{N}}|q\rangle_{in}|01\rangle_u|q\rangle_m = \frac{1}{\sqrt{N}}|q\rangle_{in}|00\rangle_u(|p\rangle_m+|q\rangle_m)+\sqrt{\frac{N-2}{N}}|q\rangle_{in}|01\rangle_u|q\rangle_m$

7. $|\Phi_7\rangle = \prod_{j=n}^{1}\text{CNOT}_{in_j u_2 m_j}|\Phi_6\rangle = \frac{1}{\sqrt{N}}|q\rangle_{in}|00\rangle_u(|p\rangle_m+|q\rangle_m)+\sqrt{\frac{N-2}{N}}|q\rangle_{in}|01\rangle_u|0\cdots0\rangle_m$.

In step 1, the pattern q in the input register is copied into the memory register, but only in the processing component. In step 2, the CNOT gate implements the sum of the binary strings encoded into the qubits of the input and the memory registers, the negation obtained by the action of the NOT is denoted by the overline. In step 3, the first utility qubit is flipped in the processing component. In step 4, we have the action of the gate (8.1.2), for $i = N-1$, on the second utility qubit controlled by the first utility qubit. In step 5, the first utility qubit in the processing component is remapped in $|0\rangle$. In step 6, the inverse operations of step 2 creates the superposition of the first two stored patterns in the memory register of the storing component. In step 7, the memory register is reset to 0 in the processing component. The system is ready for the next upload.

The seven steps illustrated above must be repeated for any pattern to be stored, using the gate $\text{C}S^{N+1-i}$ in step 4 to process the pattern $p^{(i)}$. Let us observe that after the uploading of k patterns, the overall state of the three registers is:

$$\left|\Phi_7^{(k)}\right\rangle = \frac{1}{\sqrt{N}}\sum_{i=1}^{k}\left|p^{(k)}\right\rangle_{in}|00\rangle_u\left|p^{(i)}\right\rangle_m + \sqrt{\frac{N-k}{N}}\left|p^{(k)}\right\rangle_{in}|01\rangle_u|0\cdots0\rangle_m.$$

For $k = N$, we have stored all the patterns in $\mathcal{P}$:

$$\left|\Phi_7^{(N)}\right\rangle = \frac{1}{\sqrt{N}}\sum_{i=1}^{N}\left|p^{(N)}\right\rangle_{in}|00\rangle_u\left|p^{(i)}\right\rangle_m.$$

The memory register is exactly in $|M\rangle$.

Once a collection of binary patterns of length n has been stored in a quantum memory encoded in the n-qubit state $|M\rangle$, we need an efficient retrieval procedure that allows the incomplete knowledge of the information to be retrieved. We can try to apply the Grover's algorithm for the retrieval as anticipated above. Let us consider an example presented in [VM00], let $\mathcal{P}$ be the collection of binary patterns given by:

$$\mathcal{P} = \{0000, 0011, 0110, 1001, 1100, 1111\}. \tag{8.1.4}$$

In the basis encoding, the patterns are encoded into the states of the computational basis of a 4-qubit register, with respect to this basis the memory state takes the form:

$$|M\rangle = \frac{1}{\sqrt{6}}(1,0,0,1,0,0,1,0,0,1,0,0,1,0,0,1)^T. \tag{8.1.5}$$

Suppose we need to retrieve the pattern 0110 running Grover on $|M\rangle$ as initial state. The Grover iteration $U_G U_f$ must applied $\frac{\pi}{4}\sqrt{6}$ (two) times:

$$U_f|M\rangle = \frac{1}{\sqrt{6}}(1,0,0,1,0,0,-1,0,0,1,0,0,1,0,0,1)^T,$$

$$GU_f|M\rangle = \frac{1}{2\sqrt{6}}(-1,1,1,-1,1,1,3,1,1,-1,1,1,-1,1,1,-1)^T,$$

$$U_f GU_f|M\rangle = \frac{1}{2\sqrt{6}}(-1,1,1,-1,1,1,-3,1,1,-1,1,1,-1,1,1,-1)^T,$$

$$GU_f GU_f|M\rangle = \frac{1}{8\sqrt{6}}(5,-3,-3,5,-3,-3,13,-3,-3,5,-3,-3,5,-3,-3,5)^T.$$

A measurement process in the computational basis finds the target pattern with probability:

$$\mathbb{P}(0110) = \left(\frac{13}{8\sqrt{6}}\right)^2 \simeq 44\% \tag{8.1.6}$$

The obtained probability is low because the action of the diffusion operator U_G introduces data that are not in the original storage increasing the size of the search space. Thus there are two kinds of non-desired states corresponding to stored patterns and non-stored patterns, these states present opposite phases. Grover's algorithm can be modified in such a way the non-desired states of both kinds have the same phase.

In order to define the modified Grover's algorithm for pattern recognition, we introduce an oracle operator that marks the stored patterns with a phase flip:

$$U_{\mathcal{P}}|x\rangle := (-1)^{g(x)}|x\rangle \tag{8.1.7}$$

where $g : \{0,1\}^n \to \{0,1\}$ is the indicator function:

$$g(x) = \begin{cases} 1 & x \in \mathcal{P} \\ 0 & x \notin \mathcal{P} \end{cases}.$$

Algorithm 10 is the modified Grover's algorithm used for retrieval, the action of $U_{\mathcal{P}}$ in line 4 inverts the phase of the stored patterns so that the non-desired pattern present the same phase.

Input: Collection of binary patterns $\mathcal{P}$, oracle function $f : \mathcal{P} \to \{0, 1\}$ marking $\hat{p}$.
Result: target pattern $\hat{p}$
1 Prepare $|M\rangle$;
2 $|M\rangle \leftarrow U_f |M\rangle$;
3 $|M\rangle \leftarrow U_G |M\rangle$;
4 $|M\rangle \leftarrow U_\mathcal{P} |M\rangle$;
5 $|M\rangle \leftarrow U_G |M\rangle$;
6 **Repeat** $\frac{\pi}{4}\sqrt{N} - 1$ times:
7 $\qquad |M\rangle \leftarrow U_f |M\rangle$;
8 $\qquad |M\rangle \leftarrow U_G |M\rangle$;
9 Measure

Algorithm 10: *Modified Grover's algorithm for pattern recognition.*

Let us reconsider the example of the collection (8.1.4) where the target pattern to retrieve is 0110. The state that enters in the iterative part of Algorithm 10 is:

$$GU_\mathcal{P} GU_f |M\rangle = \frac{1}{4\sqrt{6}}(1, 1, 1, 1, 1, 1, 9, 1, 1, 1, 1, 1, 1, 1, 1, 1)^T.$$

A single iteration must be performed yielding:

$$GU_\mathcal{P} GU_f |M\rangle \xrightarrow{U_f} \frac{1}{4\sqrt{6}}(1, 1, 1, 1, 1, 1, -9, 1, 1, 1, 1, 1, 1, 1, 1, 1)^T$$

$$\xrightarrow{G} \frac{1}{16\sqrt{6}}(1, 1, 1, 1, 1, 1, 39, 1, 1, 1, 1, 1, 1, 1, 1, 1)^T$$

A measurement process provides the target with probability:

$$\mathbb{P}(0110) \simeq 0.99.$$

In the case we have an incomplete pattern to retrive like 011[*], where [*] denotes the missing bit, the oracle U_f marks two states: $|0110\rangle$ encoding a stored pattern and $|0111\rangle$ encoding a non-stored pattern. The explicitly calculation, as done above, shows that the probability of returning a matching pattern is $\mathbb{P}(011[*]) \simeq 0.96$ and the probability of returning the stored pattern is $\mathbb{P}(0110) \simeq 0.78$. Therefore repeated runs are required to retrieve the desired pattern in $\mathcal{P}$.

In the general case where p_{in} is an incomplete input pattern and we need to retrieve a matching pattern from a given set $\mathcal{P}$, the number of iterations in the modified Grover's algorithm is estimated in [VM00] by the application of a general result of [Bi98].

***Proposition* 8.1.1** *Let n be the length of binary patterns, $N < 2^n$ be the number of the stored patterns, R be the number of marked states corresponding to matching (stored and unstored) patterns, k be the average amplitude of marked states in the superposition $U_G U_\mathcal{P} U_G U_f |M\rangle$ and l be the average amplitude of unmarked states in $U_G U_\mathcal{P} U_G U_f |M\rangle$. The number T of Grover's iterations in Algorithm 10 required to retrieve a matching pattern is:*

$$T = \left\lceil \frac{\pi/2 - \arctan\left(\frac{k}{l}\sqrt{\frac{R}{2^n - R}}\right)}{\arccos\left(1 - 2\frac{R}{2^n}\right)} \right\rceil .$$

The structure of the quantum associative memory that we have introduced can be summarized as follows:

1. Basis encoding of $N < 2^n$ binary patterns of length n into a n-qubit register;
2. Storing $\mathcal{P} = \{p^{(1)}, ..., p^{(N)}\}$ into $|M\rangle = \frac{1}{\sqrt{N}}\sum_i |p^{(i)}\rangle$;
3. Run the modified Grover's algorithm to retrieve a matching pattern.

Thus the time complexity of *storing* is $O(N)$ and the query complexity of retrieval is $O(\sqrt{N/R})$. Nevertheless, the obtained advantage is not a speedup with respect to a classical procedure but the exponential storage capacity of the quantum memory.

Now suppose to have a binary pattern p and a collection $\mathcal{P}$ of N binary patterns and we want to retrieve from $\mathcal{P}$ the most similar pattern to p. A method to do so is proposed in [Tr01]. We need two n-qubit registers: One storing the input pattern p and the other storing the memory state $|M\rangle$. Assuming the presence of an ancillary qubit, the initial state is:

$$|\Psi_0\rangle = \frac{1}{\sqrt{N}} \sum_{i=1}^{N} |p_1 \cdots p_n\rangle_{in} \left|p_1^{(i)} \cdots p_n^{(i)}\right\rangle_m |+\rangle, \qquad (8.1.8)$$

then a sequence of NOT and CNOT gates is applied yielding:

$$|\Psi_1\rangle = \prod_{j=1}^{n} \text{NOT}_{m_j} \text{CNOT}_{in_j m_j} |\Psi_0\rangle = \frac{1}{\sqrt{N}} \sum_{i=1}^{N} |p_1 \cdots p_n\rangle_{in} \left|d_1^{(i)} \cdots d_n^{(i)}\right\rangle_m |+\rangle, (8.1.9)$$

where $d_j^{(i)} = 1$ if and only if $p_i = p_j^{(i)}$ for all $j = 1, ..., n$. Then the state in the memory register encodes a comparison between the input pattern and the patterns in the storage. Let us consider the following Hamiltonian of the composite system made by the memory register and the ancillary qubit:

$$\mathcal{H} = \mathcal{H}_m \otimes \sigma_z \qquad (8.1.10)$$

where $\mathcal{H}_m := \sum_{j=1}^{n} \frac{\sigma_z^{(j)} + \mathbb{I}}{2}$, and $\sigma_z^{(j)}$ is the local Pauli matrix $\sigma_z = \begin{pmatrix} 1 & 0 \\ 0 & -1 \end{pmatrix}$ acting on the jth qubit of the memory register. Let us observe that $|d_1^{(i)} \cdots d_n^{(i)}\rangle_m$ is an eigenstate of $\mathcal{H}_M$ whose eigenvalue is given by the number of zeros in the string $d_1^{(i)} \cdots d_n^{(i)}$ that is nothing but the number of bits differing in p and $p^{(i)}$ that is the Hamming distance:

$$d_H(p, p^{(i)}) := \sum_{j=1}^{n} (p_j \oplus p_j^{(i)}). \tag{8.1.11}$$

The retrieval procedure is based on the time evolution of the state (8.1.9) under the action of the Hamiltonian $\mathbb{I}_{in} \otimes \mathcal{H}$, where $\mathbb{I}_{in}$ is the identity operator acting on the input register:

$$|\Psi_2\rangle = e^{i\frac{\pi}{2n}\mathbb{I}_{in} \otimes \mathcal{H}}|\Psi_1\rangle = \frac{1}{\sqrt{2N}} \sum_{i=1}^{N} e^{i\frac{\pi}{2n}d_H(p,p^{(i)})}|p_1 \cdots p_n\rangle \left|d_1^{(i)} \cdots d_n^{(i)}\right\rangle_m |0\rangle + \tag{8.1.12}$$

$$+ \frac{1}{\sqrt{2N}} \sum_{i=1}^{N} e^{-i\frac{\pi}{2n}d_H(p,p^{(i)})}|p_1 \cdots p_n\rangle \left|d_1^{(i)} \cdots d_n^{(i)}\right\rangle_m |1\rangle.$$

Applying the Hadamard gate on the ancillary qubit and the gate $\prod_{j=n}^{1} CNOT_{in_j m_j} NOT_{m_j}$ that restores the patterns $p^{(1)}, ..., p^{(N)}$ in the memory register, we obtain[1]:

$$|\Psi_3\rangle = \frac{1}{\sqrt{N}} \sum_{i=1}^{N} \cos\left[\frac{\pi}{2n}d_H(p, p^{(i)})\right] |p_1 \cdots p_n\rangle \left|p_1^{(i)} \cdots p_n^{(i)}\right\rangle_m |0\rangle + \tag{8.1.13}$$

$$+ \frac{1}{\sqrt{N}} \sum_{i=1}^{N} \sin\left[\frac{\pi}{2n}d_H(p, p^{(i)})\right] |p_1 \cdots p_n\rangle \left|p_1^{(i)} \cdots p_n^{(i)}\right\rangle_m |1\rangle.$$

A measurement process on the ancillary qubit with respect to the basis $\{|0\rangle, |1\rangle\}$ produces the outcomes according to the following probabilities:

$$\mathbb{P}(0) = \frac{1}{N} \sum_{i=1}^{N} \cos^2\left[\frac{\pi}{2n}d_H(p, p^{(i)})\right], \qquad \mathbb{P}(1) = \frac{1}{N} \sum_{i=1}^{N} \sin^2\left[\frac{\pi}{2n}d_H(p, p^{(i)})\right] \tag{8.1.14}$$

This measurement reveals an insight about the similarity between the input pattern and the stored patterns. In fact, roughly speaking, if p is very different form all

---

[1]Using the Euler formula $e^{ix} = \cos(x) + i\sin(x)$.

$p^{(1)}, ..., p^{(N)}$ then $\mathbb{P}(1)$ is high, if p is close to all $p^{(1)}, ..., p^{(N)}$ then $\mathbb{P}(1)$ is low. Thus we can fix a threshold $T \in \mathbb{N}$ so that: If T measurement processes returns 1 then p is classified as *non-recognized*; If a measurement returns 0 before the threshold is reached then p is classified as *recognized* and a pattern is retrieved from the memory. After the recognition, the retrieval can be done by a measurement on the memory register with respect to the computational basis. The outcomes are distributed according to:

$$\mathbb{P}_m(p^{(i)}) = \frac{1}{N\mathbb{P}(0)} \cos^2\left[\frac{\pi}{2n} d_H(p, p^{(i)})\right],$$

that is peaked around the patterns that are the closest to p with respect to the Hamming distance.

8.2 Pattern recognition with quantum Fourier transform

In this section we outline an interesting quantum technique of pattern recognition proposed in [Sc03]. Let us consider a $N \times M$ array given by unit cells that are either white or black like the one of Fig. 8.1. Let η be the approximately uniform density of the white cells. Assume that a fraction $\chi < 1$ of the cells in the array form a periodic structure (pattern), in a connected region, that is invariant under two translations at least into different directions.

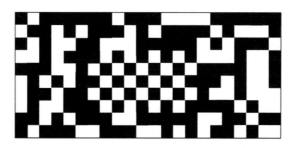

Figure 8.1. The chessboard structure in the figure can be recognized easily by a human being.

The considered pattern recognition problem can be formulated as follows: Given N, M, η, find the linear pattern in the array of cells if there is. Thus the existence of a pattern is not given a priori nor its size, shape, and orientation. Let us summarize how the quantum Fourier transform may be useful to solve this problem.

Assume that $N = 2^n, M = 2^m \gg 1$, with $n, m \in \mathbb{N}$. Let $f : \{0,1\}^n \times \{0,1\}^m \to \{0,1\}$ be the oracle function distinguishing the cells: $f(x,y) = 1$ if (x,y) are the coordinates of a white cell and $f(x,y) = 0$ otherwise. The quantum oracle defined by f is a $(n + m + 1)$-qubit gate:

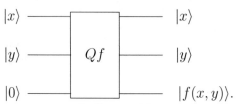

We can prepare the superposition of all possible coordinates of the cells, within the basis encoding, by means of:

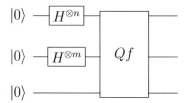

returning the state:

$$|\Psi_0\rangle = \frac{1}{\sqrt{NM}} \sum_{x=0}^{N-1} \sum_{y=0}^{M-1} |x\rangle|y\rangle|f(x,y)\rangle. \tag{8.2.1}$$

A measurement process on the third register with outcome 1 can be used to prepare the superposition of the coordinates of the all white cells:

$$|\Psi\rangle = \frac{1}{\sqrt{\eta NM}} \sum_{(x,y)\in W} |x\rangle|y\rangle \qquad W = \{\text{coordinates of the white cells}\}. \tag{8.2.2}$$

In order to apply the quantum Fourier transform we chose the more convenient representation of the $M \times N$ array in terms of the string of length $S = MN$ obtained by concatenation of the M rows. The state (8.2.2) assumes this form:

$$|\Psi\rangle = \frac{1}{\sqrt{\eta S}} \sum_{z\in Z} |z\rangle \qquad Z := \{z \in \{0,1\}^{m+n} : z \text{ coordinate of the white cells}\} \tag{8.2.3}$$

The action of the quantum Fourier transform is:

$$U_{\mathcal{F}}|\Psi\rangle = \sum_{j=0}^{S-1} \sum_{z\in Z} \frac{1}{S\sqrt{\eta}} e^{i\frac{2\pi}{S}z\cdot j}|j\rangle. \tag{8.2.4}$$

On the one hand, if there are no patterns in the considered array of cells then in the superposition (8.2.4) there are no privileged values of j, except $j = 0$. In this case a measurement process produces uniformly distributed outcomes $j > 0$. On the other hand, the presence of a pattern introduces a typical length scale inducing peaks on certain values of j that are used as indicators of the location, size and shape of the pattern.

8.3 Adiabatic pattern recognition

In section 3.4 we have introduced the general notion of adiabatic quantum computing and discussed some generalities about it. The paradigm of AQC is the preparation of a quantum system in the known ground state of an initial Hamiltonian, then an adiabatic (sufficiently slow) time evolution is performed toward the unknown ground state of a problem Hamiltonian that encodes the solution of a given problem.

Here we consider a problem Hamiltonian for pattern recognition of this general form:

$$\mathcal{H}_P = \mathcal{H}_m + \Gamma \mathcal{H}_{in} \qquad (8.3.1)$$

Where $\mathcal{H}_m$ is the *memory Hamiltonian* whose ground state encodes the considered patterns, $\mathcal{H}_{in}$ is the *input Hamiltonian* whose ground state encodes the input pattern and Γ is a weight factor.

Let $\{p^{(\mu)}\}_{\mu=1,\dots,N}$ be a collection of N binary patterns of length n to be stored, with $p_i^{(\mu)} = \pm 1$ for any $i = 1, \dots, n$. There are several options to define a good memory Hamiltonian like the following two:

$$\mathcal{H}_m = \mathbb{I} - \sum_{\mu=1}^{N} \left| p^{(\mu)} \right\rangle \left\langle p^{(\mu)} \right| \qquad (8.3.2)$$

and

$$\mathcal{H}_m = \mathbb{I} - |M\rangle\langle M| \quad \text{where} \quad |M\rangle = \frac{1}{\sqrt{N}} \sum_{\mu=1}^{N} \left| p^{(\mu)} \right\rangle. \qquad (8.3.3)$$

Let us focus on a n-qubit architecture where the Hamiltonian presents the form of a sum over the coupling terms among the qubits, by analogy with Hopfield networks:

$$\mathcal{H}_m = -\frac{1}{2} \sum_{i \neq j} w_{ij} \sigma_z^{(i)} \sigma_z^{(j)} \qquad w_{ij} := \frac{1}{n} \left[\sum_{\mu=1}^{N} p_i^{(\mu)} p_j^{(\mu)} - N\delta_{ij} \right], \qquad (8.3.4)$$

we have that the patterns $\{p^{(\mu)}\}_{\mu=1,\dots,N}$ are stored in the ground state of $\mathcal{H}_m$. The input Hamiltonian for this kind of architecture can be defined by:

$$\mathcal{H}_{in} = -\sum_{i=1}^{n} p_i^{(in)} \sigma_z^{(i)}. \qquad (8.3.5)$$

$\{|p\rangle\}_{p=1,\dots,2^n}$ are the eigenstates of $\mathcal{H}_{in}$ and the spectrum of $\mathcal{H}_{in}$ is:

$$E_{in}(p) = -n + 2d_H(p, p^{(in)}) \qquad (8.3.6)$$

where d_H is the Hamming distance. Therefore the contributions of the memory term and the input term in the Hamiltonian $\mathcal{H}_P$ entail that the states encoding stored patterns that are close to the input pattern present a high probability to be found by a measurement on the ground state of the problem Hamiltonian. More precisely, the adiabatic evolution ends in the final state:

$$|\Psi_p\rangle \simeq \frac{1}{\sqrt{N}} \sum_{\mu=1}^{N} [1 - 2\Gamma \delta d_H(\mu)] \left|p^{(\mu)}\right\rangle + O(\Gamma^2) \tag{8.3.7}$$

where $\delta d_H(\mu) = d_H(p^{(\mu)}, p) - \mathbb{E}[d_H]$ and $\mathbb{E}[d_H] = \frac{1}{N} \sum_{\mu=1}^{p} d_H(p^{(\mu)}, p)$. The initial state, that is the ground state of the initial Hamiltonian, is the uniform superposition of all the basis state:

$$|\Phi_0\rangle = \frac{1}{2^{n/2}} \sum_{k=0}^{2^n-1} |k\rangle. \tag{8.3.8}$$

The main advantage of an adiabatic algorithm for pattern recognition is that a practical realization is feasible since 2009 [Ne09]. However, there are severe shortcomings about the theoretical determination of the evolution schedule, the weight Γ, and the run time of the algorithm related to the spectral gap, as quite usual in AQC.

Chapter 9

Quantum neural networks

Artificial neural networks, or simply *neural networks*, are a class of machine learning models defined by a structure of interconnected nodes, called neurons, roughly resembling a biological brain. A neural network is characterized by the way its neurons are connected and by the weights of the connections. The weights are optimized during the training in order to accomplish a desired task. The term *quantum neural networks* may refer to artificial neurons implemented on quantum computers that can be combined in nets with a direct analogy to the classical networks but also to architectures constructed by variational quantum circuits that can be trained by backpropagation. The advantages of quantum neural networks with respect to classical counterparts are not yet clarified. However, there are some theoretical and experimental results suggesting that quantum neural networks can outperform classical networks, for instance a quantum feedforward neural network implemented in terms of a variational circuit is able to achieve a higher value of *effective dimension* (that is a figure of merit to quantify the generalization capability of a model) with respect to classical feedforward networks [Ab21]. An example of possible advantage of quantum neural networks is provided by simulations of a quantum Boltzmann machine where quantum effects enable better performances with respect to the classical version [Am18]. Anyway, proving a clear and definitive supremacy of quantum neural networks over the classical counterparts is currently an open issue which motivates the nowadays intese research activity on this topic.

In this chapter we discuss quantum implementations of the perceptron model that can be used as building block of quantum feed-forward neural networks. Then we introduce a quantum autoencoder in terms of a variational circuit, and the quantum Boltzmann machine. Moreover, we provide a general description of the quantum implementation of convolutional neural networks and generative adversarial networks.

D. Pastorello, *Concise Guide to Quantum Machine Learning*,
Machine Learning: Foundations, Methodologies, and Applications,
https://doi.org/10.1007/978-981-19-6897-6_9

9.1   Quantum perceptron

Let us consider a set of inputs X and a set of corresponding outputs Y. We need a predictive model that generalizes from data, that is, a function $f : X \to Y$ to predict the output corresponding to any new input. We assume that $X = \mathbb{R}^d$ and $Y = \mathbb{R}$. On the one hand, the predictive model can be linear, $f(\mathbf{x}, \mathbf{w}) = \mathbf{w} \cdot \mathbf{x} + w_0$ where $\mathbf{w} \in \mathbb{R}^d$ and $w_0 \in \mathbb{R}$ are the parameters of the model. On the other hand, the model can be non-linear. The *perceptron* is a non-linear model of the form:

$$f(\mathbf{x}, \mathbf{w}) = \varphi(\mathbf{w} \cdot \mathbf{x}), \tag{9.1.1}$$

where $\varphi : \mathbb{R} \to \mathbb{R}$ is a non-linear function called *activation function.*
In Fig. 9.1, the input entries are represented by nodes of a graph, called neurons. The input neurons are connected to an output neuron by weighted edges. The graph and the activation function define the model (9.1.1). The original proposal of activation function for a perceptron is the step function:

$$\varphi(a) = \left\{ \begin{array}{ll} 1 & a \geq 0 \\ -1 & a < 0 \end{array} \right. . \tag{9.1.2}$$

The training can be done by the updating rule:

$$w_j^{(t+1)} = w_j^{(t)} + \eta[y_i - \varphi(\mathbf{w} \cdot \mathbf{x})]x_{ij} , \tag{9.1.3}$$

where η is the learning rate, to obtain a convergent sequence $\{\mathbf{w}^{(t)}\}_{t>0}$ to the optimal weights. The convergence properties of the perceptron model have been studied since the 60s [No62].

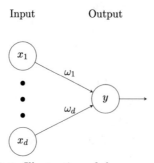

Figure 9.1. Illustration of the perceptron model.

The main shortcoming of a perceptron is that it can learn only linearly separable datasets, however many perceptrons turn out to be powerful when combined into more complex structures called *neural networks*, such as the feedforward neural network in Fig. 9.2, the perceptron itself can be considered the simplest neural network.

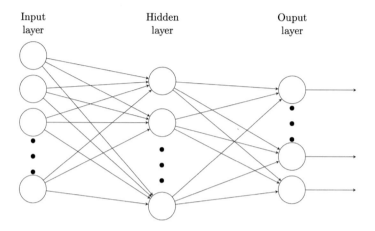

Figure 9.2. A feedforward neural network with one hidden layer.

A *feedforward neural network* is defined by multiple perceptrons organized into layers so that the neurons of each layer are only connected to the neurons of the following layer. The information is fed forward from the input layer to the output layer through the hidden layers. Feedforward neural networks present a general model function of the form:

$$f(\mathbf{x}, W_1, W_2, ...) = \cdots \varphi_2(W_2 \varphi_1(W_1 \mathbf{x})),$$

where φ_i are non-linear maps between vector spaces of appropriate dimension and W_i are the weight matrices. Popular activation functions adopted in feedforward neural networks are the Heaviside, the sigmoid, the hyperbolic tangent, the rectified linear unit (ReLU). In the case of a feedforward NN with a single hidden layer, the model function is given by:

$$f(\mathbf{x}, W_1, W_2) = \varphi_2(W_2 \varphi_1(W_1 \mathbf{x})), \tag{9.1.4}$$

and a common loss function to be minimized for the training of the network is:

$$\mathcal{L}(W_1, W_2) = \sum_{i=1}^{N} |\varphi_2(W_2 \varphi_1(W_1 \mathbf{x}_i)) - y_i|^2. \tag{9.1.5}$$

Despite its apparent simplicity, the model (9.1.4), with arbitrary activation functions, can approximate any Borel measurable function between finite dimensional spaces up to a finite precision, provided a sufficient number of nodes is available [Ho89]. Increasing the number of hidden layers, one obtain a so-called *deep neural network* that can be very performing but hard to train.

The training of a feedforward neural network is in general a non-convex, non-linear optimization problem that can be solved by *backpropagation* that is an iterative method based on the gradient descent whose update rule assumes the following form, for the case of a single hidden layer:

$$w_{ij}^{(t+1)} = w_{ij}^{(t)} - \eta \frac{\partial \mathcal{L}(W_1, W_2)}{\partial w_{ij}}. \tag{9.1.6}$$

However, since the computational complexity of each iteration within the backpropagation is polynomial in the number of connections, the convergence time can be prohibitive, moreover the search of the optimum can stop in a local minimum of $\mathcal{L}$.

Non-linearity is the key aspect of the perceptron model and the neural networks in general. Then, a quantum implementation must face the non-trivial issue of reproducing non-linearity in terms of the quantum formalism that is intrinsically linear. At first glance, quantum structures are not a natural choice for building neural networks, however interesting opportunities, like submitting superpositions of inputs and creating entanglement among neurons, offered by quantum architectures are supposed to be rather promising. Among the various proposals of quantum perceptrons [Le94, AA02, EL15, Ca17, Be19], in this section we summarize two remarkable implementations. Let us introduce a first quantum implementation of a perceptron model within the basis encoding [Sc15]. In particular, the model that we consider is given by a function $f : \{-1, 1\}^n \to \{-1, 1\}$ defined as:

$$f(\mathbf{x}, \mathbf{w}) = \varphi(\mathbf{w} \cdot \mathbf{x}) \qquad w_i \in [-1, 1] \ \forall i = 1, ..., n \tag{9.1.7}$$

where the activation function φ is the step (9.1.2). The encoding of input into qubit states is the following: $-1 \mapsto |0\rangle$, $1 \mapsto |1\rangle$.

Consider a n-qubit register (the input register) and a τ-qubit register initialized in:

$$|\Psi_0\rangle = |x_1 \cdots x_n\rangle|0 \cdots 0\rangle. \tag{9.1.8}$$

In this implementation the weights are encoded into a unitary operator acting on the input register:

$$U_{\mathbf{w}} = U_n(w_n) \cdots U_2(w_2) U_1(w_1) U_0 \tag{9.1.9}$$

where $U_i(w_i)$ acts locally on the ith qubit of the input register as:

$$\begin{pmatrix} e^{\frac{-2\pi i w_i}{2n}} & 0 \\ 0 & e^{\frac{2\pi i w_i}{2n}} \end{pmatrix} \tag{9.1.10}$$

and $U_0 = e^{i\pi}\mathbb{I}$. Therefore we have: $U_0|x_1 \cdots x_n\rangle = e^{i\pi}|x_1 \cdots x_n\rangle$ and for any $i = 1, ..., n$:

$$x_i = -1: \quad U_i(w_i)|0\rangle = \begin{pmatrix} e^{\frac{-2\pi i w_i}{2n}} & 0 \\ 0 & e^{\frac{2\pi i w_i}{2n}} \end{pmatrix} \begin{pmatrix} 1 \\ 0 \end{pmatrix} = e^{\frac{2\pi i w_i x_i}{2n}} \begin{pmatrix} 1 \\ 0 \end{pmatrix}$$

$$x_i = 1: \quad U_i(w_i)|1\rangle = \begin{pmatrix} e^{\frac{-2\pi i w_i}{2n}} & 0 \\ 0 & e^{\frac{2\pi i w_i}{2n}} \end{pmatrix} \begin{pmatrix} 0 \\ 1 \end{pmatrix} = e^{\frac{2\pi i w_i x_i}{2n}} \begin{pmatrix} 0 \\ 1 \end{pmatrix}$$

The overall action of $U_\mathbf{w}$ is:

$$U_\mathbf{w}|x_1 \cdots x_n\rangle = \exp\left[2\pi i \left(\frac{1}{2n}\mathbf{w} \cdot \mathbf{x} + \frac{1}{2}\right)\right]|x_1 \cdots x_n\rangle \equiv e^{2\pi i \phi}|\mathbf{x}\rangle.$$

The phase estimation algorithm applied to $|\Psi_0\rangle$ returns a τ-bit estimation of:

$$\phi = \frac{1}{2n}\mathbf{w} \cdot \mathbf{x} + \frac{1}{2}, \tag{9.1.11}$$

that can be directly used to evaluate the step activation function on the dot-product between the input vector and the weight vector. The phase estimation algorithm to implement the perceptron is described by the following circuit:

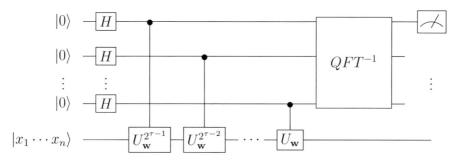

If the index register is in $|q_1 \cdots q_\tau\rangle$ then $\phi \simeq q_1\frac{1}{2^0} + \cdots + q_\tau\frac{1}{2^\tau}$. The value of q_1 decides whether $\phi \geq \frac{1}{2}$ or not, that is, whether $\mathbf{w} \cdot \mathbf{x} \geq 0$ or not. In order to evaluate the activation function, we do not need a good estimation of ϕ but just deciding if $\phi \geq \frac{1}{2}$. For this reason the number τ of qubits in the ancillary register can be small, e.g. for $\tau = 2$ the probability of success is 85%, for $\tau = 8$ the probability of success is 99.9% [Sc15]. Unfortunately, the time complexity of the quantum perceptron implementation is $O(n\tau)$ and it is comparable to that of a classical perceptron implementation. Moreover, in the basis encoding there is no a quantum space efficiency.

In order to implement a quantum perceptron based on amplitude encoding, we follow the idea introduced in [Ta19]. Let us consider binary-valued input and weight vectors $\mathbf{x}, \mathbf{w} \in \mathbb{R}^d$ such that $x_i, w_i \in \{-1, 1\} \ \forall i = 0, ...d-1$, that can be stored into a N-qubit register, $(N = \log d)$.

$$|\psi_\mathbf{x}\rangle = \frac{1}{\sqrt{d}} \sum_{i=0}^{d-1} x_i |i\rangle \qquad |\psi_\mathbf{w}\rangle = \frac{1}{\sqrt{d}} \sum_{i=0}^{d-1} w_i |i\rangle. \qquad (9.1.12)$$

The implementation of a quantum perceptron within the amplitude encoding is done following these steps:

1. Initialize the register in the idle state $|0\rangle^{\otimes N}$.

2. Prepare the input state $|\psi_\mathbf{x}\rangle = U_\mathbf{x}|0\rangle^{\otimes N}$.

3. Apply $U_\mathbf{w}$ that satisfies $U_\mathbf{w}|\psi_\mathbf{w}\rangle = |1\rangle^{\otimes N} = |d-1\rangle$.

4. Prepare an ancillary qubit in $|0\rangle_a$.

5. Action of a $N+1$-qubit CNOT on the ancilla controlled by the register.

6. Measure the ancilla.

After the first three steps, the state stored in the register is:

$$|\phi_{\mathbf{w},\mathbf{x}}\rangle = U_\mathbf{w} U_\mathbf{x}|0\rangle^{\otimes N} = U_\mathbf{w}|\psi_\mathbf{x}\rangle = \sum_{i=0}^{d-1} c_i |i\rangle, \qquad (9.1.13)$$

where c_i are the amplitudes with respect to the computational basis. Let us observe the following interesting fact:

$$\langle\psi_\mathbf{w}|\psi_\mathbf{x}\rangle = \frac{1}{d}\mathbf{w} \cdot \mathbf{x} = \langle\psi_\mathbf{w}|U_\mathbf{w}^\dagger U_\mathbf{w}\psi_\mathbf{x}\rangle = \langle d-1|U_\mathbf{w}\psi_\mathbf{x}\rangle = c_{d-1}.$$

Thus the amplitude of the state $|d-1\rangle = |1\cdots1\rangle$ in the superposition $|\phi_{\mathbf{w},\mathbf{x}}\rangle$ corresponds to the dot product of the input vector with the weights vector. After the preparation in $|0\rangle$ of an ancillary qubit (step 4), the action of the CNOT of step 5 is:

$$\mathsf{CNOT}_{N,a}|\phi_{\mathbf{w},\mathbf{x}}\rangle|0\rangle_a = \sum_{i=1}^{d-2} c_i|i\rangle|0\rangle_a + c_{d-1}|d-1\rangle|1\rangle_a$$

Finally, a measurement on the ancilla (step 6) provides the outcome 1 with probability:

$$\mathbb{P}_a(1) = |c_{d-1}|^2 = d^{-2}|\mathbf{w} \cdot \mathbf{x}|^2. \tag{9.1.14}$$

If the state of the ancilla $|0\rangle$ represents the inactive neuron and $|1\rangle$ represents the active neuron then the probability (9.1.14) describes the non-linear activation of the output neuron. In general, the complexity to implement this quantum perceptron is $O(d)$ using *quantum hypergraph states* [Ta19].

9.2 Quantum feedforward neural networks

In the previous section we have introduced two quantum implementations of a perceptron. In both cases the output is obtained measuring an ancillary qubit. If we want to use a quantum perceptron as the building block of a feedforward neural network, an option is realizing a *hybrid* neural network where the classical output of a neuron, produced by a measurement process, can be used to feed the next layer after a quantum re-encoding. Therefore, a hybrid neural network presents layers made by quantum neurons that can be individually implemented on a quantum hardware and a measurement process for any neuron between the layers. On the other hand, we can realize a *coherent* quantum neural network where quantum information is directly feed-forwarded through the network without measurement processes between the layers. In this case neurons can be in a superposition of the states of active neuron and inactive neuron. Unlike the hybrid neural network, the coherent version requires all nodes to be implemented simultaneously on a quantum register, thus making the quantum computation more demanding. However, coherent quantum neural networks reduce the necessity to store and process classical data between the layers and offer more opportunities for use on quantum processors [Ta20].

A quantum feedforward neural network can be provided in terms of layers of quantum gates acting on a n-qubit register, instead of weights and biases there are the parameters of these gates, the action of the network is given by a variational quantum circuit. For instance, the architecture discussed in [FN18] consists in a quantum processor, acting on a $n+1$ qubits register, which implements a variational circuit described by the parametric unitary operator $U(\theta)$. A measurement on the last qubit, prepared in $|1\rangle$, is used to assign one of two possible labels to the input state. The training for the classification task is based on the loss function:

$$\mathcal{L}(\theta) = -\sum_{i=1}^{N} y_i \langle x_i \, 1|U^\dagger(\theta)\sigma_z^{(n+1)}U(\theta)|x_i \, 1\rangle, \tag{9.2.1}$$

where $\{(x_i, y_i)\}_{i=1,...,N}$ is the training set with $y_i \in \{-1, 1\}$ for any $i = 1, ..., N$ and $\sigma_z^{(n+1)}$ is the operator on $(\mathbb{C}^2)^{\otimes(n+1)}$ acting as the Pauli matrix σ_z on the last qubit and as the identity on the n-qubit register. Multiple runs of the quantum circuit are required to compute the expectation values and the stochastic gradient descent can be used for the minimization of (9.2.1).

Generally speaking, the training of a quantum feedforward neural network requires classical optimization methods and efficient updates of the gates within variational quantum circuits. The general structure of a quantum feedforward neural network implemented in terms of a variational circuit is the following:

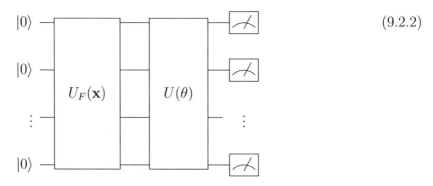

$$(9.2.2)$$

where a classical input $\mathbf{x} \in \mathbb{R}^m$ is encoded into a n-qubit quantum register by applying the feature map $|\psi_\mathbf{x}\rangle = U_F(\mathbf{x})|0\rangle^{\otimes n}$. Then a variational circuit $U(\theta)$ realizes the parametric quantum process which must be trained and a measurement extracts the output. An example of feature map for the encoding of $\mathbf{x} \in \mathbb{R}^2$ into a qubit pair is:

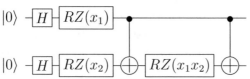

where the gate $RZ(x_i)$ is a rotation on the Bloch sphere around the z-axis by an angle which depends on the i component of the input vector. The optimization of the parameters in the variational quantum circuit $U(\theta)$ can be carried on calculating its partial derivatives with respect to the parameters and performing the gradient descent. In many interesting cases, the *parameter-shift rule* is a strategy to estimates gradients of variational quantum circuits [Mi18, Sc19]. Given a quantum observable A (typically a tensor product of Pauli matrices σ_z), the feature map and the circuit $U(\theta)$ defines the function of a parametric model:

$$f(\mathbf{x}, \theta) := \langle\psi_\mathbf{x}|U^\dagger(\theta)A\,U(\theta)|\psi_\mathbf{x}\rangle. \qquad (9.2.3)$$

$U(\theta)$ can be decomposed into a sequence of single-parameter gates and we can assume that the parameter θ_i affects only the gate $U_i(\theta_i)$ (if θ_i affects more than one gate we can apply the product rule in the differentiation that we are going to consider). Thus we have:

$$U(\theta) = V U_i(\theta_i) W, \qquad (9.2.4)$$

where V and W are the operators representing all the other gates forming the circuit which depend on the other parameters. The partial derivative of f is:

$$\frac{\partial}{\partial \theta_i} f = \frac{\partial}{\partial \theta_i} \langle \psi | U_i^{\dagger}(\theta_i) B \, U_i(\theta_i) | \psi \rangle$$

$$= \langle \psi | \frac{\partial}{\partial \theta_i} U_i^{\dagger}(\theta_i) B \, U_i(\theta_i) | \psi \rangle + \langle \psi | U_i^{\dagger}(\theta_i) B \frac{\partial}{\partial \theta_i} U_i(\theta_i) | \psi \rangle,$$

where $B = V^{\dagger} A V$ and $|\psi\rangle = W|\psi\rangle_{\mathbf{x}}$. An interesting issue is whether the partial derivatives of a parametric circuit can be implemented as unitary evolutions to estimate the gradient. A particular but relevant case is given by:

$$U_i(\theta_i) = e^{-i\theta_i H_i}, \qquad (9.2.5)$$

where the self-adjoint generator H_i has two distinct eigenvalues $\pm r$. This requirement is satisfied by any 1-qubit gate for instance. The partial derivative of f is:

$$\frac{\partial}{\partial \theta_i} f = \langle \psi' | B(-iH_i) | \psi' \rangle + \langle \psi' | (iH_i) B | \psi' \rangle \quad \text{with} \quad |\psi'\rangle = U_i(\theta_i)|\psi\rangle. \qquad (9.2.6)$$

The following technical lemma turns out to be a useful tool for computing the gradient of variational circuits.

Lemma 9.2.1 *Let B, C, D be three linear operators on a Hilbert space. Then:*

$$C^{\dagger} B D + D^{\dagger} B C = \frac{1}{2}[(C + D)^{\dagger} B(C + D) - (C - D)^{\dagger} B(C - D)].$$

Proof. Direct calculation. $\qquad \square$

Applying Lemma 9.2.1 for $C = \mathbb{I}$ and $D = -ir^{-1}H_i$, we get:

$$\frac{\partial}{\partial \theta_i} f = \frac{r}{2} \left(\langle \psi' | (\mathbb{I} - ir^{-1}H_i)^{\dagger} B(\mathbb{I} - ir^{-1}H_i) | \psi' \rangle - \right.$$

$$\left. - \langle \psi' | (\mathbb{I} + ir^{-1}H_i)^{\dagger} B(\mathbb{I} + ir^{-1}H_i) | \psi' \rangle \right). \qquad (9.2.7)$$

Since H_i has at most two distinct eigenvalues $\pm r$, the sine and cosine parts of the Taylor series of $U_i(\theta_i)$ take the following form (as explicitly shown in [Sc19]):

$$U_i(\theta_i) = \mathbb{I}\cos(r\theta_i) - ir^{-1}H_i\sin(r\theta_i),$$

as a consequence:

$$U_i\left(\frac{\pi}{4r}\right) = \frac{1}{\sqrt{2}}(\mathbb{I} - ir^{-1}H_i). \tag{9.2.8}$$

Inserting (9.2.8) in (9.2.7), we obtain the following:

$$\frac{\partial}{\partial\theta_i}f = r\left(\langle\psi|U_i^\dagger(\theta_i + s)BU_i(\theta_i + s)|\psi\rangle - \langle\psi|U_i^\dagger(\theta_i - s)BU_i(\theta_i - s)|\psi\rangle\right)$$

$$= r(f(\theta_i + s) - f(\theta_i - s)) \quad \text{where} \quad s = \frac{4\pi}{r}. \tag{9.2.9}$$

Equation (9.2.9) is called parameter-shift rule and implies that the partial derivatives of a variational circuit can be computed by using the same variational architecture. In other words, the same circuit $U(\theta)$ is used for the computation of the model function (9.2.3) and of its gradient as well. In this sense, the parameter update is performed classically but the variational approach to the optimization is based on a quantum calculation of the gradient. As an example, consider the parametric 1-qubit gate defined by:

$$U_i(\theta_i) = e^{-i\theta_i\sigma_\mu}, \tag{9.2.10}$$

where σ_μ is a Pauli matrix then $U_i(\theta_i)$ is a rotation around the μ-axis of the Bloch sphere by an angle θ_i. The eigenvalues of σ_μ are ± 1 then the partial derivative of f with respect to θ_i can be calculated by means of the parameter-shift rule with $s = \pi/4$. The variational quantum circuit considered in [Ab21], used to show that a quantum neural network can achieve a better effective dimension with respect to classical feedforward networks, is made up by parametric Pauli rotations and CNOT gates then it can be differentiated by the parameter-shift rule.

9.3 Quantum autoencoder for data compression

In this section we define a quantum autoencoder for data compression following the proposal of [Ro17]. A classical autoencoder is a feed-forward neural network where the input layer and the output layer present the same number of neurons and the hidden layers have fewer neurons (Fig. 9.3).

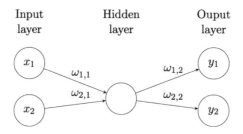

Figure 9.3. An elementary autoencoder represented as a feed-forward neural network.

The training is aimed to find the optimal weights to transfer information, as faithfully as possible, from the input layer to the output layer. When the autoencoder is perfectly trained then it acts as the identity. Given a set of inputs $\{\mathbf{x}_i\}_i$, the network can be trained minimizing the loss function:

$$\mathcal{L}(\mathbf{w}) = \sum_i \| \mathbf{x}_i - \mathbf{y}_i^{(\mathbf{w})} \|^2 . \tag{9.3.1}$$

Once completed the training, the model has learnt to compress and decompress data that are close to the training set. The compressed data are taken directly from the layer with fewest neurons. Data can be decompressed by the processing with the remaining layers.

In order to implement a quantum autoencoder, let us consider a n-qubit register A and a k-qubit register B. Let $\{|\psi_i\rangle_{AB}\}_i$ be a training set of pure states, the goal is to learn a compression procedure of $n + k$ qubits into n qubits. Let $\{U(\mathbf{w})\}_{\mathbf{w}\in\mathbb{R}^m}$ be a parametrized family of $(n + k)$-qubit gates and $|\varphi\rangle_{B'}$ be a reference state of k qubits stored in a copy B' of the register B. Let us consider the following circuit:

$$\tag{9.3.2}$$

that returns the final state $\rho_i^{(\mathbf{w})} = \text{tr}_{B'}\left[\rho_{out,i}^{(\mathbf{w})}\right]$ in the register $A + B$. The training of the circuit is done by the minimization of the function:

$$\mathcal{L}(\mathbf{w}) = -\sum_i \mathcal{F}(|\psi_i\rangle_{AB}, \rho_i^{(\mathbf{w})}), \tag{9.3.3}$$

where $\mathcal{F}(|\psi\rangle, \rho) := \langle\psi|\rho\,\psi\rangle$ is the fidelity. Assume that we have set the parameters $\mathbf{w}$ such that the output state $\rho_{B',i}^{(\mathbf{w})}$ in the regiser B' coincides to the input state in B':

$$\rho_{B',i}^{(\mathbf{w})} := \mathrm{tr}_{AB}\left(\rho_{out,i}^{(\mathbf{w})}\right) = |\varphi\rangle\langle\varphi|_{B'}, \tag{9.3.4}$$

then we have that in the register B, after the action of $U(\mathbf{w})$ and before the SWAP, there is the state $|\varphi\rangle_B$, obtained by the reduced evolution:

$$\mathrm{tr}_A[U(\mathbf{w})|\psi_i\rangle\langle\psi_i|_{AB}U^\dagger(\mathbf{w})] = |\varphi\rangle\langle\varphi|_B. \tag{9.3.5}$$

By unitarity of $U(\mathbf{w})$, we have that the evolved state $U(\mathbf{w})|\psi_i\rangle\langle\psi_i|_{AB}U^\dagger(\mathbf{w})$ is a pure state. Since its partial trace is pure as well then $U(\mathbf{w})|\psi_i\rangle\langle\psi_i|_{AB}U^\dagger(\mathbf{w})$ is non-entangled, so the corresponding state vector can be written as a product:

$$U(\mathbf{w})|\psi_i\rangle = |\psi_i^c\rangle_A|\varphi\rangle_B, \tag{9.3.6}$$

where $|\psi_i^c\rangle_A$ is a n-qubit compressed version of $|\psi_i\rangle_{AB}$. Therefore the circuit (9.3.2) acts as the identity on the input states:

$$|\psi_i\rangle_{AB}|\varphi\rangle_{B'} \xrightarrow{U(\mathbf{w})\otimes\mathbb{I}_{B'}} |\psi_i^c\rangle_A|\varphi\rangle_B|\varphi\rangle_{B'} \xrightarrow{\mathbb{I}_A\otimes SWAP} |\psi_i^c\rangle_A|\varphi\rangle_B|\varphi\rangle_{B'} \xrightarrow{U^\dagger(\mathbf{w})\otimes\mathbb{I}_{B'}} |\psi_i\rangle_{AB}|\varphi\rangle_{B'}.$$

We have that if the parameter vectors $\mathbf{w}$ satisfies the requirement (9.3.4) then $\mathbf{w} = \mathrm{argmin}\mathcal{L}$ and represents a procedure of compression and decompression of $n + k$-qubit states into n-qubit states learnt from $\{|\psi_i\rangle_{AB}\}_i$. Thus the training of the quantum autoencoder can be done by the minimization of:

$$\mathcal{L}_1(\mathbf{w}) = -\sum_i \mathcal{F}\left(|\varphi\rangle_{B'}, \rho_{B',i}^{(\mathbf{w})}\right). \tag{9.3.7}$$

Let us remark that the quantum states of the training set can be unknown because the training is based only on the knowledge of the reference state $|\varphi\rangle_{B'}$ and only on a measurement processes over the register B'. The structure of a quantum autoencoder for data compression is based on the general notion of variational quantum circuit where the parameters update is done classically. For instance the gradient of $\mathcal{L}_1$ can be calculated using existing techniques for evaluating gradients of variational functions and exploiting the established infrastructure for automatic differentiation of classical functions. Otherwise, one can apply the automatic differentiation of variational quantum circuits via the parameter-shift rule described in section 9.2.

9.4 Quantum Boltzmann machine

A Boltzmann machine (BM) is a stochastic recurrent neural network where the neurons corresponds to binary units divided into *visible* and *hidden* nodes representing variables $z_i \in \{-1, +1\}$. It is a well-known machine learning technique serving as the basis of deep learning models [Sa09].

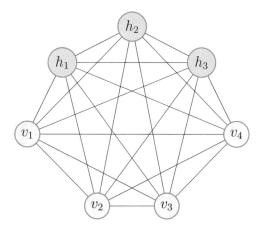

Figure 9.4 A fully connected Boltzmann machine.

In a BM, each node of the network is attached to a bias term w_i and the weighted connections w_{ij} represent the couplings among the nodes so that the network is characterized by the energy function of an Ising model:

$$E(\mathbf{z}) = -\sum_{i=1}^{n} w_i z_i - \sum_{i<j} w_{ij} z_i z_j, \qquad (9.4.1)$$

the weights $\{w_i, w_{ij}\}$ are tuned by training. The vector $\mathbf{z}$ represents the configuration of the n units, the configuration vectors of the hidden and visible units are denoted by $\mathbf{h}$ and $\mathbf{v}$ respectively. According to the description of the Ising model in statistical mechanics, at the equilibrium, the probability that the visible units are in the state $\mathbf{v}$ is given by the Boltzmann distribution:

$$\mathbb{P}(\mathbf{v}) = Z^{-1} \sum_{\mathbf{h}} e^{-E(\mathbf{z})} \qquad \text{with} \qquad Z = \sum_{\mathbf{z}} e^{-E(\mathbf{z})}. \qquad (9.4.2)$$

The probability distribution $\mathbb{P}$ is calculated as a marginal distribution taking the sum $\sum_{\mathbf{h}}$ over the hidden variables. The goal of the training of a BM is to determine the weights w so that (9.4.2) is as close as possible to an input probability distribution

$\mathbb{P}^{data}$ defined by the training set. The loss function to minimize can be defined by the average negative log-likelihood:

$$\mathcal{L}(\mathbb{P}^{data}, \mathbb{P}) := -\sum_{\mathbf{v}} \mathbb{P}^{data}(\mathbf{v}) \ln \mathbb{P}(\mathbf{v}). \tag{9.4.3}$$

The minimization can be done by the gradient descent. The weights are iteratively changed by a step in the direction opposite to the gradient:

$$\delta w = -\eta \partial_w \mathcal{L}, \tag{9.4.4}$$

where $\eta > 0$ is the learning rate. The gradient of $\mathcal{L}$ where $\mathbb{P}$ is given by (9.4.2) is:

$$\partial_w \mathcal{L} = \sum_{\mathbf{v}} \mathbb{P}^{data}(\mathbf{v}) \left(\frac{\sum_{\mathbf{h}} \partial_w E(\mathbf{z}) e^{-E(\mathbf{z})}}{\sum_{\mathbf{h}} e^{-E(\mathbf{z})}} - \frac{\sum_{\mathbf{z}} \partial_w E(\mathbf{z}) e^{-E(\mathbf{z})}}{\sum_{\mathbf{z}} e^{-E(\mathbf{z})}} \right). \tag{9.4.5}$$

The corresponding parameter variations are:

$$\delta w_i = \eta \left(\sum_{\mathbf{v}} \mathbb{P}(\mathbf{v})^{data} \frac{\sum_{\mathbf{h}} z_i \, e^{-E(\mathbf{z})}}{\sum_{\mathbf{h}} e^{-E(\mathbf{z})}} - Z^{-1} \sum_{\mathbf{z}} z_i e^{-E(\mathbf{z})} \right) \tag{9.4.6}$$

$$= \eta \left(\sum_{\mathbf{v}} \mathbb{P}^{data}(\mathbf{v}) \langle z_i \rangle_{\mathbf{v}} - \langle z_i \rangle \right),$$

$$\delta w_{ij} = \eta \left(\sum_{\mathbf{v}} \mathbb{P}(\mathbf{v})^{data} \frac{\sum_{\mathbf{h}} z_i z_j \, e^{-E(\mathbf{z})}}{\sum_{\mathbf{h}} e^{-E(\mathbf{z})}} - Z^{-1} \sum_{\mathbf{z}} z_i z_j e^{-E(\mathbf{z})} \right) \tag{9.4.7}$$

$$= \eta \left(\sum_{\mathbf{v}} \mathbb{P}^{data}(\mathbf{v}) \langle z_i z_j \rangle_{\mathbf{v}} - \langle z_i z_j \rangle \right),$$

where $\langle \ \rangle$ denotes the Boltzmann average and $\langle \ \rangle_{\mathbf{v}}$ denotes the Boltzmann average with fixed visible variables to the data. Both the averages can be efficiently determined by sampling the units of the network.

The quantum Boltzmann machine (QBM) is a model where the units of the BM are implemented by qubits [Am18] with Ising Hamiltonian:

$$\mathcal{H} = -\sum_i w_i \sigma_z^{(i)} - \sum_{ij} w_{ij} \sigma_z^{(i)} \sigma_z^{(j)}. \tag{9.4.8}$$

The eigenstates of $\mathcal{H}$, forming the computational basis, correspond to all the 2^n configurations of the hidden and visible units. The density operator:

$$\rho = Z^{-1}e^{-\mathcal{H}} \quad \text{with} \quad Z = \text{tr}(e^{-\mathcal{H}}), \tag{9.4.9}$$

represents the equilibrium state of the quantum Ising model. The eigenvalues of ρ provide the Boltzmann distribution of the 2^n configurations of the network units. The measurement process with respect to the computational basis on the visible units is $\{|\mathbf{v}\rangle\langle\mathbf{v}| \otimes \mathbb{I_h}\}_{\mathbf{v}}$, where $\mathbb{I_h}$ is the identity acting on the Hilbert space of the hidden qubits, and the marginal distribution over the visible units can be calculated by:

$$\mathbb{P}(\mathbf{v}) = \text{tr}(\Lambda_{\mathbf{v}}\rho) \quad \text{with} \quad \Lambda_{\mathbf{v}} := |\mathbf{v}\rangle\langle\mathbf{v}| \otimes \mathbb{I_h}. \tag{9.4.10}$$

The quantum model described so far is nothing but a re-formulation of the classical BM. The quantum nature of the network can be exploited considering a transverse Ising Hamiltonian:

$$\mathcal{H} = -\sum_i \Gamma_i \sigma_x^{(i)} - \sum_i w_i \sigma_z^{(i)} - \sum_{ij} w_{ij}\sigma_z^{(i)}\sigma_z^{(j)}, \tag{9.4.11}$$

so the eigenstates of $\mathcal{H}$ are superpositions of the states encoding the configurations of the network units. The average negative log-likelihood that we consider for training the QBM is:

$$\mathcal{L} = -\sum_{\mathbf{v}} \mathbb{P}^{data}(\mathbf{v}) \ln\left[\frac{\text{tr}(\Lambda_{\mathbf{v}}e^{-\mathcal{H}})}{\text{tr}(e^{-\mathcal{H}})}\right], \tag{9.4.12}$$

with gradient:

$$\partial_\theta \mathcal{L} = \sum_{\mathbf{v}} \mathbb{P}^{data}(\mathbf{v})\left(\frac{\text{tr}(\Lambda_{\mathbf{v}}\partial_\theta e^{-\mathcal{H}})}{\text{tr}(\Lambda_{\mathbf{v}}e^{-\mathcal{H}})} - \frac{\text{tr}(\partial_\theta e^{-\mathcal{H}})}{\text{tr}(e^{-\mathcal{H}})}\right). \tag{9.4.13}$$

Since $\mathcal{H}$ and $\partial_\theta\mathcal{H}$ are non-commuting operators, we have that $\partial_\theta e^{-\mathcal{H}} \neq -e^{-\mathcal{H}}\partial_\theta\mathcal{H}$. However the chain rule can be recovered in the weak form [Am18]:

$$\text{tr}(\partial_\theta e^{-\mathcal{H}}) = -\text{tr}(\partial_\theta\mathcal{H}\,e^{-\mathcal{H}}). \tag{9.4.14}$$

Therefore the term in (9.4.13) with free visible units can be estimated as a Boltzmann average by sampling:

$$\frac{\text{tr}(\partial_\theta e^{-\mathcal{H}})}{\text{tr}(e^{-\mathcal{H}})} = -\frac{\text{tr}(\partial_\theta\mathcal{H}\,e^{-\mathcal{H}})}{\text{tr}(e^{-\mathcal{H}})} = -\text{tr}(\partial_\theta\mathcal{H}\rho) = -\langle\partial_\theta\mathcal{H}\rangle_\rho. \tag{9.4.15}$$

On the other hand, the term with fixed visible units, the so-called *clamped* term, cannot be estimated by sampling. In order to train the model with a gradient descent carried on by sampling, QBM can be re-formulated as the *bound-based* QBM. By the Golden-Thompson inequality:

$$\text{tr}(e^A e^B) \geq \text{tr}(e^{A+B}), \tag{9.4.16}$$

that is true for any pair of hermitian matrices A and B, we can determine a lower bound of the marginal distribution (9.4.10) on the visible units:

$$\mathbb{P}(\mathbf{v}) = Z^{-1}\text{tr}(e^{-\mathcal{H}}e^{\ln \Lambda_{\mathbf{v}}}) \geq Z^{-1}\text{tr}(e^{-\mathcal{H}+\ln \Lambda_{\mathbf{v}}}). \tag{9.4.17}$$

Defining a new Hamiltonian by:

$$\mathcal{H}_{\mathbf{v}} := \mathcal{H} - \log \Lambda_{\mathbf{v}}, \tag{9.4.18}$$

we can determine an upper bound of the average negative log-likelihood:

$$\mathcal{L} \leq \widetilde{\mathcal{L}} = -\sum_{\mathbf{v}} \mathbb{P}^{data}(\mathbf{v}) \log \frac{\text{tr}(e^{-\mathcal{H}_{\mathbf{v}}})}{\text{tr}(e^{-\mathcal{H}})}. \tag{9.4.19}$$

In order to train the QBM, instead of minimizing $\mathcal{L}$, we minimize its upper bound $\widetilde{\mathcal{L}}$. The gradient is:

$$\partial_\theta \widetilde{\mathcal{L}} = \sum_{\mathbf{v}} \mathbb{P}^{data}(\mathbf{v}) \left(\frac{\text{tr}(\partial_\theta \mathcal{H}_{\mathbf{v}}\, e^{-\mathcal{H}_{\mathbf{v}}})}{\text{tr}(e^{-\mathcal{H}_{\mathbf{v}}})} - \frac{\text{tr}(\partial_\theta \mathcal{H}\, e^{-\mathcal{H}})}{\text{tr}(e^{-\mathcal{H}})} \right), \tag{9.4.20}$$

thus the parameter variations can be expressed as follows:

$$\delta\theta_i = \eta \left(\sum_{\mathbf{v}} \mathbb{P}^{data}(\mathbf{v}) \langle \sigma_z^{(i)} \rangle_{\mathbf{v}} - \langle \sigma_z^{(i)} \rangle \right) \tag{9.4.21}$$

$$\delta\theta_{ij} = \eta \left(\sum_{\mathbf{v}} \mathbb{P}^{data}(\mathbf{v}) \langle \sigma_z^{(i)}\sigma_z^{(j)} \rangle_{\mathbf{v}} - \langle \sigma_z^{(i)}\sigma_z^{(j)} \rangle \right) \tag{9.4.22}$$

where $\langle\ \rangle$ is the Boltzmann average with Hamiltonian $\mathcal{H}$ and $\langle\ \rangle_{\mathbf{v}}$ is the Boltzmann average with the clamped Hamiltonian $\mathcal{H}_{\mathbf{v}}$.

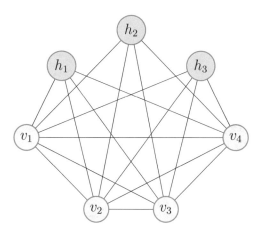

Figure 9.5 A semi-restricted Boltzmann machine.

Within the bound-based formulation, the training of a QBM requires a sampling with clamped Hamiltonian to calculate $\langle\ \rangle_{\mathbf{v}}$ for any input data instance, then it may be very consuming for large datasets. A less demanding training procedure can be done defining a *restricted* QBM obtained removing the connections among the hidden units. The connections among visible units are allowed then it is a semi-restricted BM indeed (Fig. 9.5). In this case, the Hamiltonian (9.4.18) takes the form:

$$\mathcal{H}_{\mathbf{v}} = -\sum_i \Gamma_i \sigma_x^{(i)} + w_i^{eff}(\mathbf{v})\sigma_z^{(i)} \quad \text{with} \quad w_i^{eff}(\mathbf{v}) = w_i + \sum_{j\ visible} w_{ij}v_j, \quad (9.4.23)$$

allowing the exact computation of the clamped Boltzmann averages:

$$\langle\sigma_z^{(i)}\rangle_{\mathbf{v}} = \frac{w_i^{eff}}{\sqrt{\Gamma_i^2 + (w_i^{eff})^2}} \tanh\sqrt{\Gamma_i^2 + (w_i^{eff})^2}. \quad (9.4.24)$$

In order to test whether a QBM is better than a classical BM in learning data thanks to the quantum fluctuations introduced by the transverse field, one must to perform empirical validations of the model with a quantum machine, like a quantum annealer, or simulating the qubits network. QBM training through minimizing both the negative log-likelihood and its upper bound within the bound-based formulation must be compared to the results with classical BM training. For small-size datasets, there are examples where QBM learns the data distribution better than classical BM. In this respect, let us report some results of a simulation which highlight a possible advantage of the QBM [Am18]. Consider the synthetic training set constructed starting from a random selection of a center point $\mathbf{s}^k \in \{-1, +1\}^N$ and the Bernoulli distribution:

$$\mathbb{P}_k(\mathbf{v}) = p^{N-d(\mathbf{v},\mathbf{s}^k)}(1 - p)^{d(\mathbf{v},\mathbf{s}^k)} \qquad p \in [0, 1], \tag{9.4.25}$$

where d is the Hamming distance. The data distribution is defined by:

$$\mathbb{P}^{data}(\mathbf{v}) = \frac{1}{M} \sum_{k=1}^{M} \mathbb{P}_k(\mathbf{v}). \tag{9.4.26}$$

Consider the Kullback-Liebler divergence to quantify the quality of learning:

$$KL(\mathbb{P}^{data}, \mathbb{P}) = \sum_{\mathbf{v}} \mathbb{P}^{data}(\mathbf{v}) \log \frac{\mathbb{P}^{data}(\mathbf{v})}{\mathbb{P}(\mathbf{v})}. \tag{9.4.27}$$

Let us consider a fully visible, fully connected model with $N = 10$ and restrict the Hamiltonian (9.4.11) of the QBM by setting $\Gamma_i = \Gamma$ for all i. Then the training parameters are Γ, θ_i and θ_{ij}. In the case of he bound-based QBM (bQBM) Γ is set to 2 and the training parameters are θ_i and θ_{ij}. In the simulation presented in [Am18], the expectation values in the gradients of the log-likelihood are computed exactly.

	Limit KL	Convergence
BM	≈ 0.62	10 iterations
QBM	≈ 0.42	30 iterations
bQBM	≈ 0.50	20 iterations

Table 9.1. Trainings of BM, QBM and bQBM to learn the distribution (9.4.26) [Am18].

The comparison of the trainings of BM, QBM and bQBM are reported in table 9.1. The training is done iteratively by means of the Broyde-Fletcher-Goldfarb-Shanno algorithm that performs line search along the gradient. The limit KL indicates the Kullback-Liebler divergence between the learned distribution and the input distribution once the optimization of log-likelihood converged. The training of the QBM is slower but it provides a better result.

9.5 Quantum convolutional neural networks

Convolutional neural networks (CNNs) are architectures used in image recognition. An input image is mathematically described by a matrix of pixel values, each value corresponds to the pixel brightness in the considered image (Fig. 9.6).

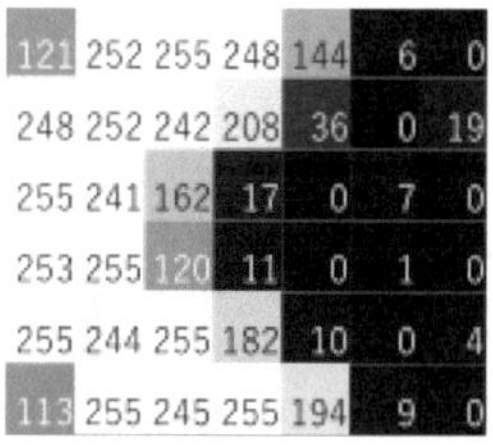

Figure 9.6. A 6×7 image represented by its pixel values ranging from 0 to 255.

A CNN is a sequence of interspersed *convolutional layers* and *pooling layers*, any layer produces an intermediate matrix of pixel values, called *feature map*. The convolutional layers compute the new pixel values $x_{ij}^{(l)}$ as linear combinations of nearby pixels from the previous feature map:

$$x_{ij}^{(l)} = \sum_{a,b=1}^{W} w_{a,b} x_{i+a,j+b}^{(l-1)}, \qquad (9.5.1)$$

where the weights $w_{a,b}$ form a $W \times W$ matrix. The pooling layers reduce the dimension of the received feature map, for example by extracting patches from the input feature maps and taking the maximum value in each patch discarding the other values. Pooling layers are followed by the action of a nonlinear activation function. Once the size of the feature map is small enough the *fully connected layer* computes a function which depends on all remaining pixels and returns the output. The training of a CNN optimizes the weights and the fully connected function while the number of layers and W are fixed a priori.

The CNN architecture can be reconsidered in terms of quantum circuits to construct a quantum convolutional neural network (QCNN) [CCL19]. The input is an unknown quantum state ρ_{in}, a convolutional layer is implemented by means of the repeated application of a few-qubit quantum gate U_i in parallel. A pooling layer reduces the size of the input measuring a fraction of the qubits and acting with a gate V_i on the nearby qubit controlled by the outcomes of the measurements. The measurement processes and the conditional action of V_j implement the nonlinear activation function. The fully connected layer is represented by a gate F on the remaining qubits. The output is read measuring a fixed number of qubits. The following quantum circuit is an example of a simple QCNN with two convolutional layers and two pooling layers, the fully connected layer involves only two qubits with a final measurement process on a single qubit:

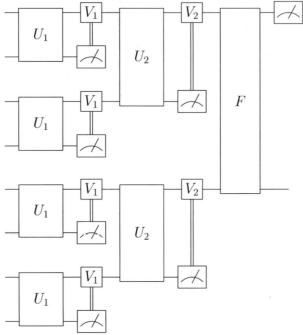

The number of convolutional and pooling layers is fixed a priori and the gates U_i, V_i, F are learned. Once defined a loss function, for example the mean-squared error in the output with respect to the training set, one can optimize the parameters of the variational circuit. To process a n-qubit input state, the number of parameters to learn is $O(\log n)$. This efficiency makes QCNNs interesting quantum architectures to investigate. Two possible applications of QCNNs are related to recognition of unknown quantum states and to quantum error correction [CCL19].

9.6   Quantum generative adversarial networks

Generative adversarial networks (GANs) are schemes where two neural networks play the role of two agents in a *zero-sum game* for generating new data distributed with the same statistic of the given training data [Go14]. The generative network, called *generator*, maps latent variables z, distributed according to a prior distribution p, into fake data $x = G(z)$. The other network, called *discriminator*, must distinguish the real data from the data generated by G. The GAN training can be formulated in terms of cross-entropy:

$$\min_G \max_D \left\{ \mathbb{E}_{x \sim p_{data}(x)}[\log(D(x))] + \mathbb{E}_{z \sim p(z)}[\log(1 - D(G(z)))] \right\}, \qquad (9.6.1)$$

where $\mathbb{E}_{x \sim p_{data}(x)}$ is the expectation value over the distribution p_{data} of the training set and $\mathbb{E}_{z \sim p(z)}$ is the expectation value over the latent variable distribution p. G and D are the functions implemented by the generative and discriminative networks, in particular $D(x)$ is the probability that x comes from the real data distribution. According to the model (9.6.1) the discriminator attempts to maximize both the log-probability of predicting that a real x is genuine and the log-probability of predicting that a generated $x = G(z)$ is fake. The final result provided by the GAN is a generator model that produces samples from the observed distribution by sampling the prior distribution p over the latent space.

In this section, we discuss a proposal for a quantum GAN assuming the availability of universal fault-tolerant quantum computers [Ll18], but also a hybrid quantum-classical approach to GANs [Ro21]. Firstly, let us consider a generator and a discriminator as abstract objects: the first one is a source of quantum states that emits with a statistic and the second one is a measurement strategy. Assume that data are quantum: the training set is formed by an ensemble of quantum states described by a density matrix σ on the Hilbert space H. The generator provides an ensemble of quantum states distributed according to a density matrix ρ. The first step is to train the discriminator to distinguish between σ and ρ. The second step is to train the generator in producing fake states, fixing the measurement strategy. As we have discussed in section 7.5, the minimum error measurement to discriminate between two given states σ and ρ can be constructed out as the Helstrom measurement. However, the discriminator has not the knowledge to implement the optimal measurement then it guess a measurement $\{P_{true}, P_{false}\}$. The probability that the outcome of the measurement is *true* on a state that is selected from the data ensemble is:

$$\mathbb{P}(data|true) = \text{tr}(P_{true}\sigma). \tag{9.6.2}$$

The probability that the discriminator recognizes a generated fake state as true is:

$$\mathbb{P}(gen|true) = \text{tr}(P_{true}\rho). \tag{9.6.3}$$

The discriminator tunes the available measurements, applying the gradient descent to $\mathbb{P}(data|true)$ for finding the optimal distinguishing measurement. The task of the generator is adjusting the state ρ to maximize the probability (9.6.3), tuning the parameters of a quantum circuit for instance. As in the classical case, in this adversarial game the unique Nash equilibrium occurs when $\rho = \sigma$ and $\mathbb{P}(data|true) = \mathbb{P}(data|true) = 1/2$ [Ll18].

An alternative approach to quantum GANs is given by a hybrid quantum-classical architecture in terms of a *variational quantum generator* (VQG) [Ro21].

Consider a classical data source described by an unknown distribution. The VQG is designed to generate fake samples that reproduce the data distribution, it includes two variational quantum circuits: an encoding circuit $R(z)$ over a r-qubit register and a generator circuit $G(\theta_g)$ over a n-qubit circuit, with $n \geq r$.

The encoding circuit $R(z)$ take the latent variable $z \in \mathbb{R}^N$, with prior distribution p, as parameters and prepares the state:

$$R(z)|0\rangle^{\otimes r} = |\phi(z)\rangle, \tag{9.6.4}$$

generating a manifold of states $\{|\phi(z)\rangle\}_{z \in \mathbb{R}^N}$ which plays the role of a quantum latent space. The action of the generator circuit:

$$G(\theta_g)(|0\rangle^{\otimes (n-r)}|\phi(z)\rangle) = |\psi(z,\theta)\rangle, \tag{9.6.5}$$

and a subsequent measurement process map the latent manifold to the observed data. The state $|\psi(z,\theta_g)\rangle$ is converted in classical information taking the expectation values of Pauli matrices over M qubits and obtaining the sample:

$$P = (\langle P_1\rangle_{\psi(z,\theta_g)}, ..., \langle P_M\rangle_{\psi(z,\theta_g)}) \in \mathbb{R}^M, \tag{9.6.6}$$

where $P_i \in \{\mathbb{I}, \sigma_x, \sigma_y, \sigma_z\}$. Finally, P is classically processed by a function that outputs the generated data:

$$x_{gen} = f_g(P). \tag{9.6.7}$$

The general scheme of the VQG, implementing the overall function $F_G(z, \theta_g)$, can be described by the following quantum circuit (the measurements of the Pauli operators are taken over M qubits of the n-qubit register):

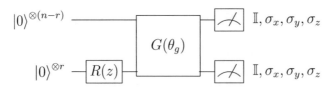

Let $x \mapsto F_D(x, \theta_d)$ be a discriminator function, parametrized by θ_d, which receives a sample x and returns the probability that x originates from the real distribution. The discriminator can be a classical feed-forward neural network, but also a *variational quantum classifier* (VQC). The quantum discriminator is formed by an encoding circuit $E(x)$ mapping the received data point x into a quantum state, a variational circuit $D(\theta_d)$, with parameters θ_d, and a measurement process. Since the classical output of the quantum discriminator is a probability value, the final measurement is simply performed on a single qubit with respect to the computational basis. The discriminator provides the probability that x coming from the real distribution as:

$$F_D(x, \theta_d) = \frac{1 + \langle \sigma_z^{(p)} \rangle_D}{2}, \tag{9.6.8}$$

where p is the label of the designated qubit and:

$$\langle \sigma_z^{(p)} \rangle_D = \langle 0 | E^\dagger(x) D^\dagger(\theta_d) \sigma_z^{(p)} D(\theta_d) E(x) | 0 \rangle. \tag{9.6.9}$$

The training of the variational hybrid GAN can be formulated as the following problem:

$$\min_{\theta_g} \max_{\theta_d} \left\{ \mathbb{E}_{x \sim p_{data}(x)} [\log(F_D(x, \theta_d))] + \mathbb{E}_{z \sim p(z)} [\log(1 - F_D(F_G(z, \theta_g), \theta_d))] \right\} \tag{9.6.10}$$

that can be solved by a gradient based method. Once established the general architecture above, one can perform experiments with variational circuits, implementing VQG and VQD, like the numerical simulations with 2-qubit generator and a 3-qubit discriminator presented in [Ro21].

As usual in the field of quantum neural networks, a significant open issue is whether a quantum GAN can offer an advantage with respect to classical models of generative learning. The peculiarity of VQG approach is the possibility of combining variational circuits, which are growing in size within the novel quantum architectures and are the core of the NISQ computation, and standard neural networks realizing the classical post-processing in the generator or implementing a classical discriminator.

Chapter 10

Concluding remarks

In this volume on the foundations of quantum machine learning (QML), we started from fundamentals of quantum mechanics and quantum computation to list some applications of well known quantum algorithms to devise machine learning schemes. In several cases, a quantum speedup over classical machine learning algorithms can be theoretically proved, however an effective implementation of QML algorithms is currently complicated on real quantum hardware. Nevertheless, for some QML schemes the situation is particularly tricky, also from the theoretical point of view, like quantum perceptrons and neural networks, where the advantage of a quantum implementation is not yet clarified. In general, we can conclude that a rather common shortcoming of several QML models is the need of a training phase performed updating parameters by means of classical optimization techniques. So, many QML proposals offer a quantum *execution* of a model but often the training remains a classical procedure.

For small datasets and low-dimensional feature spaces, some of the described algorithms can be easily implemented and tested on real quantum machines with few qubits like the IBM backends [IBM]. However, in general, the QML algorithms overviewed in this volume require the availability of quantum computers that are universal, fault-tolerant, large-scaled. Unfortunately, the current and near-term noisy intermediate-scale quantum devices (NISQ devices [Pr18]) fail these three requirements. The existing quantum machines present hardware architectures with a hundred of qubits that are little connected and not error-corrected. There are also specific-purpose quantum computers with a high number of qubits, like the quantum annealers, which have a narrower application potential. Moreover, most QML algorithms are based on the availability of a QRAM that allows an efficient quantum state preparation. The QRAM architecture is very expansive in terms of space resources and it is far off from an operating implementation. For these reasons, an interesting direction of investigation is the development of QML schemes and quantum procedures of data storing/retrieval that can be effectively implemented on NISQ devices. In this respect hybrid quantum-classical algorithms like the well-known *Quantum Approximate Optimisation Algorithm* [Fa14] and the *Variational Eigensolver* [Pe14] represent promising tools to devise machine learning schemes for near-term quantum machines [Sc18]. Also quantum-inspired machine learning, of which we give an example in section 7.5, is a prominent topic with a current impact where quantum formalism is relevant in devising new machine learning algorithms for classical hardware.

Indeed QML is not only *quantum-enhanced machine learning*, that is, the translation of celebrated machine learning procedures into the language of quantum computing applying old quantum algorithms as subroutines. From a broader viewpoint, QML represents a novel approach to data encoding and data processing that stimulates the development of completely new learning procedures without classical counterparts. The learning capability of quantum machines is probably little understood nowadays and perhaps it may be used to develop new quantum algorithms in terms of "quantum meta-learning". As a matter of fact, the huge resources theoretically offered by quantum computers seem to be partly frustrated by the difficulty of designing effective quantum algorithms, as evidenced by their relative shortage. Thus, a desirable direction of scientific investigation concerns the development of learning mechanisms to devise new efficient quantum algorithms. Some recent hybrid quantum-classical approaches to meta-learning have been proposed, like [Ve19, Wi21, Li21, PB19, Pa21] for instance.

Two antipodal attitudes towards QML are commonly served up: the *enthusiastic viewpoint* and the *dismissive viewpoint*. The first one is roughly motivated by the exponential speedup in computation that can be achieved with quantum computers, at a first glance this seems to be enough to enable a revolution in machine learning and artificial intelligence. The dismissive viewpoint is mainly related to the fact that large-scale fault-tolerant universal quantum computers are not yet available then many QML schemes cannot be realistically implemented. Furthermore, there is the lack of generalized procedures to efficiently encode large amounts of classical data into quantum hardware. These opposite points of view are founded on true premises but both draw conclusions which are probably too severe. On the one hand, the possibilities offered by the paradigms of quantum computing and the availability of quantum hardware definitely deserve the huge and intense research activity that is currently carried on. On the other hand, we must be honest in our claims paying much attention on what can be really done with existing and near-term quantum machines. However, the value of a scientific result cannot be uniquely measured by its immediate spendability in concrete applications.

In conclusion, QML is not merely a collection of applications of quantum computing to machine learning schemes but it should be considered a new approach to learning from data. A general definition of QML could be: "the study of the learning capability of controlled quantum systems and the relative applications to data science", in this sense QML may include also possible research activities of philosophical relevance on the connection between the abstract concept of *learning* and the *quantum nature* of the world.

Bibliography

[Ab21] A. Abbas et al. *The power of quantum neural networks* Nature Computational Science 1, 403-409 (2021)

[Ah07] D. Aharonov, W. van Dam, J. Kempe, Z. Landau, S. Loyd, O. Regev *Adiabatic qunatum computing is equivalent to standard quantum computation* SIAM Journal of Computing, Vol. 37, Issue 1, p. 166-194 (2007)

[Al16] S. Al-Ananzi, H. AlMahmoud, I. Al-Turaiki *Finding similar documents using different clustering techniques* Procedia Computer Science 82 (2016)

[Aï07] W. Aïmeur, G. Brassard, S. Gambs *Quantum clustering algorithms* ICML '07: Proceedings of the 24th international conference on Machine learning (2007)

[Aï13] E. Aïmeur, G. Brassard, S. Gambs, *Quantum speed-up for unsupervised learning* Mach. Learn. 90(2), 261-287 (2013)

[Am18] M. H. Amin et al. *Quantum Boltzmann Machine* Phys. Rev. X 8, 021050 (2018)

[AA02] M. Andrecut and M. Ali *A quantum perceptron* International Journal of Modern Physics B 16, 639 (2002)

[Ar15] S. Arunachalam et al. *On the robustness of bucket brigade quantum RAM* New Journal of Physics, Vol. 17, No. 12, Pp. 123010 (2015)

[ADR82] A. Aspect, J. Dalibard, G.Roger *Experimental test of Bell's inequlities using time-varying analyzers* Phys. Rev. Lett. 49, 1804-1807 (1982).

[Ba20] A. A. Basheer, S. K. Goyal *Quantum k-nearest neighbor machine learning algorithm* preprint (2020)

[Ba18] C. Bauckhage, C. Ojeda, R. Sifa, S. Wrobel *Adiabatic Quantum Computing for Kernel k=2 Means Clustering* Proceedings of the Conference LWDA 2018 21-32 (2018)

[Ba20] J. Bausch *Recurrent Quantum Neural Networks* Advances in Neural Information Processing Systems 33 (NeurIPS 2020)

[BBHT98] M. Boyer, G. Brassard, P. Hoyer, and A. Tapp *Tight bounds on quantum searching* Fortschr. Phys., vol. 4, no. 5, pp. 493-505 (1998)

[Be20] K. Beer et al. *Training deep quantum neural networks* Nature Communications 11, 808 (2020)

[Be64] J. S. Bell *On Einstein Podolski Rosen paradox* Physics 1, 195-200 (1964)

[Be66] J. S. Bell *On the problem of hidden variables in quantum mechanics* Reviews in Modern Physics, vol 38 (1966)

[Be19] F. Benatti, S. Mancini, and S. Mangini, *Continuous variable quantum perceptron* International Journal of Quantum Information 17, 1941009 (2019).

[Be80] P. Benioff *The computer as a physical system: A microscopic quantum mechanical Hamiltonian model of computers as represented by Turing machines* Journal of Statistical Physics, vol. 22, n. 5, pp. 563-591 (1980)

[Bi98] D. Biron, O. Biham, E. Biham, M. Grassl, A.L. Daniel *Generalized Grover search algorithm for arbitrary initial amplitude distribution* Proceedings of the First NASA International Conference on Quantum Computation and Quantum Communications, pp. 140-147 (1998).

[BF28] M. Born, V. A. Fock *Beweis des Adiabatensatzes* Zeitschrift für Physik A 51, 3-4 (1928)

[BH97] G. Brassard, P. Hoyer *An exact quantum polynomial-time algorithm for Simon's problem* Proceedings of Fifth Israeli Symposium on Theory of Computing and Systems. IEEE Computer Society Press: 12-23 (1997)

[BL74] W.S. Brainerd, L.H. Landweber *Theory of Computation* (Wiley 1974)

[Br01] L.D. Brown, T.T. Cai, A. DasGupta *Interval Estimation for a Binomial Proportion* Statist. Sci. 16 (2) 101 - 133 (2001)

[Bu01] H. Buhrman, R. Cleve, J. Watrous, R. de Wolf *Quantum Fingerprinting* Phys. Rev. Lett. 87, 167902 (2001)

[Ca17] Y. Cao, G. G. Guerreschi, and A. Aspuru-Guzik, *Quantum neuron: an elementary building block for machine learning on quantum computers* arXiv:1711.11240 (2017).

[Ce20] J. Cervantes, F. Garcia-Lamont, L. RodrÃguez-Mazahua, A. Lopez *A comprehensive survey on support vector machine classification: Applications, challenges and trends* Neurocomputing, Volume 408, **2020**, 189–215.

[CN00] I. Chuang, M.A. Nielsen *Quantum Computation and Quantum Information* Cambridge University Press (2000).

[Cl98] R. Cleve, A. Ekert, C. Macchiavello *Quantum algorithms revisited* Proceedings of the Royal Society of London A, vol. 454, n. 1969, (1998)

[CCL19] I. Cong, S. Choi, M. D. Lukin *Quantum Convolutional Neural Networks* Nature Physics (2019)

[Co94] D. Coppersmith *An approximate Fourier transform useful in quantum factoring* Technical Report RC19642, IBM (1994)

[Co09] T. H. Cormen et al. *Introduction to Algorithms*, 3rd ed., MIT Press 2009

[CV95] C. Cortes, V. Vapnik *Support-vector networks* Mach Learn **20**, 273-297 (1995).

[Cr93] D. Crevier *AI: The Tumultuous Search for Artificial Intelligence*. New York, NY: BasicBooks (1993)

[DC05] A. Das, B.K. Chakrabarti *Quantum annealing and related optimization methods* Springer Lecture Notes in Physics 679 (2005)

[De85] D. Deutsch *Quantum theory, the Church-Turing principle and the universal quantum computer* Proceedings of the Royal Society A, vol. 400, n. 1818 (1985)

[DJ92] D. Deutsch, R. Jozsa *Rapid solutions of problems by quantum computation* in Proceedings of the Royal Society of London A, vol. 439, n. 1907 (1992)

[DH96] C. Dürr, P. Høyer *A quantum algorithm for finding the minimum* arXiv:quant-ph/9607014 (1996)

[Do08] . D. Dong, C. Chen, H. Li, T.-J. Tarn *Quantum Reinforcement Learning* IEEE Transactions on Systems Man and Cybernetics Part B: Cybernetics, Vol. 38, No. 5, pp.1207-1220, 2008

[DW21a] D-Wave URL https://www.dwavesys.com [online, last access on August 15, 2022]

[DW21b] D-Wave Systems Inc., *Minor embedding*, `https://docs.dwavesys.com/docs/latest/c_gs_3.html#minor-embedding` [online, last access on August 15 2022]

[DW21c] D-Wave Systems Inc., *Minor embedding example*, `https://docs.dwavesys.com/docs/latest/c_gs_7.html#getting-started-embedding` [online, last access on August 15 2022]

[EPR35] A. Einstein, B. Podolski, N. Rosen *Can Quantum-Mechanical descritpion of physical reality be considered complete?* Phys. Rev. 47, 777 (1935)

[EL15] T. Espinosa-Ortega and T. Liew *Perceptrons with Hebbian learning based on wave ensembles in spatially patterned potentials* Physical Review Letters 114, 118101 (2015).

[FGG00] E. Farhi, J. Goldstone, S. Gutmann, M. Sipser *Quantum computation by adiabatic evolution* quant-ph report n. 0001106 (2000)

[Fa14] E. Farhi, J. Goldstone, S. Gutmann *A Quantum Approximate Optimization Algorithm* arXiv:1411.4028 (2014)

[FN18] E. Farhi and H. Neven *Classification with Quantum Neural Networks on Near Term Processors* arXiv:1802.06002 (2018)

[Fe82] R. Feynman *Simulating Physics with Computers* International Journal of Theoretical Physics, vol. 21, n. 6/7, pp. 467-488 (1982)

[FA17] C. Focil-Arias, J. Ziiniga, G. Sidorov, I. Batyrshin, A. Gelbukh *A tweets classifier based on cosine similarity* CEUR Workshop Proceedings (2017)

[FR99] L. Fortnow, J. Rogers *Complexity limitations on Quantum computation* J. Comput. Syst. Sci. 59 (2): 240-252 (1999)

[Gi08] V. Giovannetti, S. Lloyd, L. Maccone *Quantum random access memory* Phys. Rev. Lett. 100, 160501 (2008)

[Giu21] Giuntini, R., Freytes H., Park, D.K., Blank, C., Holik, F., Chow, K.L., Sergioli, G. Quantum state discrimination for supervised classification. *arXiv:quanth-ph/2104.00971v1* 2021.

[Go14] I. J. Goodfellow et al. *Generative adversarial nets* NIPS'14: Proceedings of the 27th International Conference on Neural Information Processing Systems, vol. 2, pages 2672-2680 (2014)

[Gr96] L.K. Grover *A fast quantum mechanical algorithm for database search* Proceedings of the twenty-eighth annual ACM symposium on Theory of Computing, STOC '96 (1996)

[Ha09] A.W. Harrow, A. Hassidim, S. Lloyd *Quantum algorithm for linear systems of equations* Phys. Rev. Lett. 103 (15) (2009) 150502.

[He69] Helstrom, C.W. *Quantum detection and estimation theory.* J. Stat. Phys., 1(2), 231–252. (1969)

[He15] B. Hensen et al. *Loophole-free Bell inequality violation using electron spins separated by 1.3 kilometres* Nature 526, 682-686 (2015)

[Ho82] J. J. Hopfield *Neural networks and physical systems with emergent collective computational abilities* in Proceedings of the National Academy of Sciences of the USA, vol. 79 no. 8 pp. 2554 2558 (1982).

[Ho89] K. Hornik, M. Stinchcombe, H. White. *Multilayer feedforward networks are universal approximators.* Neural Networks, 2(5):359-366 (1989)

[Ho12] K. Hornik, I. Feinerer, M. Kober, C. Buchta *Spherial k-means clustering* Journal of Statistical Software vol. 50, n. 10 (2012)

[Hu20] H.-Y. Huang et al. *Power of data in quantum machine learning* arXiv preprint arXiv:2011.01938, (2020)

[IBM] IBM Quantum URL https://www.ibm.com/quantum-computing [online, last access on August 15 2022]

[JRS07] S. Jensen, M. Ruskai, R. Seiler *Bounds for the adiabatic approximation with applications to quantum computation* J Math Phys 48:102111 (2007)

[Je21] S. Jerbi et al. *Quantum Enhancements for Deep Reinforcement Learning in Large Spaces* PRX Quantum 2, 010328 (2021)

[KN98] T. Kadowaki, H. Nishimori *Quantum annealing in the transverse Ising model* Phys. Rev. E 58 (1998)

[KN98] T. Kadowaki and H. Nishimori (Nov 1998), *Quantum annealing in the transverse Ising model,* Phys. Rev. E, Vol.58, pp. 5355-5363.

[Ka50] T. Kato *On the Adiabatic Theorem of Quantum Mechanics* J. Phys. Soc. JPN 5, 6 (1950)

[Ko14] G. Kochenberger, J. Hao, F. Glover, M. Lewis, Z. Lü, H. Wang, Y. Wang *The unconstrained binary quadratic programming problem: A survey*, Journal of Combinatorial Optimization, Vol.28, Num.1, pp. 58-81 (2014)

[KM40] M. Krein, D. Milman *On extreme points of regular convex sets* Studia Mathematica 9 133-138 (1940)

[LP21] R. Leporini, D. Pastorello *Support vector machines with quantum state discrimination* Quantum Rep. 2021, 3(3), 482-499

[LP22] R. Leporini, D. Pastorello *An efficient geometric approach to quantum-inspired classifications* Scientific Reports vol. 12 (1), (2022)

[Le94] M. Lewenstein, *Quantum perceptrons* Journal of Modern Optics 41, 2491 (1994)

[Li21] Y. Liang et al. *A hybrid quantum-classical neural network with deep residual learning* Neural Networks, vol. 143, pp. 133-147 (2021)

[Li99] G. Lindblad *A general no-cloning theorem* Lett. Math. Phys. 47(2), 189-196 (1999)

[Ll13] S. Lloyd, M. Mohseni, P. Rebentrost *Quantum algorithms for supervised and unsupervised machine learning* (2013) arXiv:1307.0411v2

[Ll14] S. Lloyd, M. Mohseni, P. Rebentrost *Quantum principal component analysis* Nature Physics 10, 631 (2014)

[Ll18] S. Lloyd, C. Weedbrook *Quantum generative adversarial learning* Phys. Rev. Lett. 121, 040502 (2018)

[Ma21] Y. Ma, H. Song, J. Zhang, *Quantum Algorithm for K-Nearest Neighbors Classification Based on the Categorical Tensor Network States* International Journal of Theoretical Physics, vol. 60, n. 3, 1164-1174 (2021)

[Mc14] C. C. McGeoch *Adiabatic quantum computation and quantum annealing* Synthesis lectures on quantum computing, Morgan & Claypool Publishers (2014)

[Mi18] K. Mitarai, M. Nagoro, M. Kitagawa, K. Fujii *Quantum circuit learning* Phys. Rev. A 98, 032309 (2018)

[Mo18] N. Moll et al. *Quantum optimization using variational algorithms on near-term quantum devices* Quantum Sci. Technol. 3 030503 (2018)

[MN08] S. Morita, H. Nishimori *Mathematical foundation of quantum annealing* Journal of Mathematical Physics 49 (2008)

[Mü90] B. Müller, J. Reinhardt *Neural Networks* Springer-Verlag, Berlin (1990)

[Ne09] R. Neigovzen et al. *Quantum pattern recognition with liquid state nuclear magnetic resonance* Phys. Rev. A 79, 042321 (2009)

[No62] A.B.J. Novikoff *On convergence proofs on perceptrons* In: Proceedings of the Symposium on the Mathematical Theory of Automata, vol. 12, pp. 615-622 (1962)

[PPR19] D.K. Park, F. Petruccione, JK.K. Rhee *Circuit-Based Quantum Random Access Memory for Classical Data* Sci Rep 9, 3949 (2019)

[Pa19] D. Pastorello *Entanglement, CP-Maps and Quantum Communications* In: Quantum Physics and Geometry, Lecture Notes of the Unione Matematica Italiana 2E, Ballico et al. (eds.), 5 (2019)

[PB19] D. Pastorello, E. Blanzieri. *Quantum Annealing Learning Search for solving QUBO problems.* Quantum Information Processing 18: 303 (2019)

[PB21] D. Pastorello, E. Blanzieri *A quantum binary classifier based on cosine similarity* to be appear in proceedings of IEEE conference on quantum engineering (2021)

[Pa21] D. Pastorello, E. Blanzieri, V. Cavecchia *Learning adiabatic quantum algorithms over optimization problems* Quantum Machine Intelligence vol. 3, n. 2 (2021).

[Pe04] A. Peres, D. Terno, *Quantum Information and Relativity Theory* Rev. Mod. Phys., vol. 76, n. 1, pp. 93-123 (2004)

[Pe14] A. Peruzzo et al. *A variational eigenvalue solver on a photonic quantum processor.* Nat. Commun. 5 (2014)

[Pr18] J. Preskill *Quantum computing in the NISQ era and beyond* Quantum, vol. 2, p. 79 (2018)

[Qiskit] Qiskit URL https://qiskit.org [online, last access on August 15, 2022]

[Re14] P. Rebentrost, M. Mohseni, S. Lloyd *Quantum support vector machine for big data classification* Phys. Rev. Lett. 113, 130503 (2014)

[Rig] Rigetti, 2022. URL https://www.rigetti.com/what-we-build [online, last access on 15 Aug 2022]

[Ro17] J. Romero et al. *Quantum autoencoders for efficient compression of quantum data* Quantum Sci. Technol. 2 045001 (2017)

[Ro21] J. Romero, A. Aspuru-Guzik *Variational quantum generators: Generative adversarial quantum machine learning for continuous distributions* Adv. Quantum Technol., 4, 2000003 (2021)

[Ro58] F. Rosenblatt *The perceptron: A probabilistic model for information storage and organization in the brain* Psychological Review. 65 (6): 386-408 (1958).

[Ru18] Y. Ruan et al. *Quantum algorithm for K-nearest neighbors classification based on the metric of Hamming distance* Int. J. Theor. Phys. 56:3496-3507 (2017)

[Sg21] V. Saggio et al. *Experimental quantum speed-up in reinforcement learning agents* Nature volume 591, pages 229-233 (2021)

[Sa09] R. Salakhutdinov, G. E. Hinton *Deep Boltzmann machines*, AISTATS 2009.

[Sa19] A. Sarma, R. Chatterjee, K. Gili, T. Yu *Quantum Unsupervised and Supervised Learning on Superconducting Processors* Quant. Inf. Comput. 20 (2020)

[Sc03] E. Schützhold *Pattern recognition on a quantum computer* Phys. Rev. A 67, 062311 (2003)

[Sc14] M. Schuld, I. Sinayskiy, F. Petruccione *Quantum Computing for Pattern Classification.* Pacific Rim International Conference on Artificial Intelligence, 208-220 (2014).

[Sc15] M. Schuld, I. Sinayskiy, F. Petruccione *Simulating a perceptron on a quantum computer* Physics Letters A, 379, pp. 660-663 (2015)

[Sc17] M. Schuld et al. *Implementing a distance-based classifier with a quantum interference circuit* EPL 119 60002 (2017)

[Sc18] M. Schuld, F. Petruccione *Supervised Learning with Quantum Computers* Springer Nature (2018).

[Sc19] M. Schuld, V. Bergholm, C. Gogolin, J. Izaac, N. Killoran *Evaluating analytic gradients on quantum hardware* Phys. Rev. A 99, 032331 (2019)

[Sc22] M. Schuld, N. Killoran *Is quantum advantage the right goal for quantum machine learning?* arXiv:2203.01340 [quant-ph] (2022)

[Se17] G. Sergioli, G.M. Bosyk, E. Santucci, R. Giuntini *A quantum-inspired version of the classification problem.* Int. J. Theor. Phys., *56(12)*, 3880–3888 (2017).

[Se19] G. Sergioli, R. Giuntini, H. Freytes *A new quantum approach to binary classification.* PLoS ONE 14(5) (2019)

[Sh94] P.W. Shor *Algorithms for quantum computation: discrete logarithms and factoring* Proceedings 35th Annual Symposium on Foundations of Computer Science. IEEE Comput. Soc. (1994)

[Ta19] F. Tacchino et al. *An Artificial Neuron Implemented on an Actual Quantum Processor* npj Quantum Information (2019)

[Ta20] F. Tacchino et al. *Quantum implementation of an artificial feed-forward neural network* Quantum Science and Technology 5, 044010 (2020)

[Tr01] C. A. Trugenberger *Probabilistic quantum memories* Phys. Rev. Lett. 87, 067901 (2001)

[Tu50] A. Turing *Computing Machinery and Intelligence.* Mind. 59 (236): 433-460 (1950).

[VM00] D. Ventura, T. Martinez *Quantum associative memories* Information Sciences 124 (2000) 273-296

[Ve19] G. Verdon et al. *Learning to learn with quantum neural networks via classical neural networks* Preprint, arXiv: 1907.05415 (2019)

[Vo19] H.Vojtěch et al. *Supervised learning with quantum-enhanced feature spaces.* Nature, 567(7747) : 209 (2019)

[We98] G. Weihs, et al. *Violation of Bell's inequality under strict Einstein locality conditions* Phys. Rev. Lett. 81, 5039 (1998)

[Wi20] D. Willsch, M. Willsch, H. De Raedt, K. Michielsen *Support vector machines on the D-Wave quantum annealer* Comput. Phys. Commun. 248, 107006 (2020)

[Wi21] M. Wilson et al. *Optimizing quantum heuristic with meta-learning* Quantum Machine Intelligence 3 (2021)

[Wi14] P. Wittek *Quantum Machine Learning* Academic Press (2014).

[Wi15] N. Wiebe, A. Kapoor, K. Svore *Quantum Algorithms for Nearest-Neighbor Methods for Supervised and Unsupervised Learning* Quantum Information and Computation 15(3,4): 0318-0358 (2015)

[WZ82] W. Wootters, W. Zurek *A Single Quantum Cannot be Cloned* Nature. 299 (5886): 802-803 (1982)

[Ya93] A. Yao *Quantum circuit complexity.* 34th Annual Symposium on Foundations of Computer Science pp 352-361. (1993)

[Zh17] S. Zhang et al. *Learning k for kNN classification* ACM Transactions on Intelligent Systems and Technology (2017)